# essentials

*essentials* liefern aktuelles Wissen in konzentrierter Form. Die Essenz dessen, worauf es als „State-of-the-Art" in der gegenwärtigen Fachdiskussion oder in der Praxis ankommt. *essentials* informieren schnell, unkompliziert und verständlich

• als Einführung in ein aktuelles Thema aus Ihrem Fachgebiet
• als Einstieg in ein für Sie noch unbekanntes Themenfeld
• als Einblick, um zum Thema mitreden zu können

Die Bücher in elektronischer und gedruckter Form bringen das Fachwissen von Springerautorinnen kompakt zur Darstellung. Sie sind besonders für die Nutzung als eBook auf Tablet-PCs, eBook-Readern und Smartphones geeignet. essentials sind Wissensbausteine aus den Wirtschafts-, Sozial- und Geisteswissenschaften, aus Technik und Naturwissenschaften sowie aus Medizin, Psychologie und Gesundheitsberufen. Von renommierten Autorinnen aller Springer- Verlagsmarken.

Patric U. B. Vogel • Günter A. Schaub

# New Infectious Diseases in Germany and Europe

Springer

Patric U. B. Vogel  
Cuxhaven, Germany

Günter A. Schaub  
Ruhr-Universität Bochum  
Bochum, Germany

ISSN 2197-6708        ISSN 2197-6716   (electronic)  
essentials  
ISBN 978-3-658-41825-0      ISBN 978-3-658-41826-7   (eBook)  
https://doi.org/10.1007/978-3-658-41826-7

This book is a translation of the original German edition „Neue Infektionskrankheiten in Deutschland und Europa" by Vogel, Patric U. B., published by Springer Fachmedien Wiesbaden GmbH in 2021. The translation was done with the help of artificial intelligence (machine translation by the service DeepL.com). A subsequent human revision was done primarily in terms of content, so that the book will read stylistically differently from a conventional translation. Springer Nature works continuously to further the development of tools for the production of books and on the related technologies to support the authors.

This Springer imprint is published by the registered company Springer Fachmedien Wiesbaden GmbH, part of Springer Nature.  
The registered company address is: Abraham-Lincoln-Str. 46, 65189 Wiesbaden, Germany

# What You Can Find in This *Essential*

- An introduction to new infectious diseases in Europe
- The presentation of the health consequences
- An overview of the settlement of new mosquito species that are potential vectors of tropical and subtropical pathogens
- An introduction to factors favouring the spread of vectors and viruses
- An overview of various infectious diseases, from African swine fever to COVID-19 and West Nile fever

# Contents

# Introduction

1

In the last 15 years, various new **infectious diseases** have reached Germany and Europe or have the potential to become endemic. In the time before that, there were always new infectious diseases, but with longer time intervals. Individual diseases have managed to manifest themselves undetected in our country, such as AIDS in the 1980s, while we have been largely spared from others, such as the **SARS pandemic** of 2002/2003. Many of the new dangers are favoured by increasing globalisation and climate change. For example, global warming has caused the average temperature in Germany to rise by about 1.4 °C over the last 135 years. Although not the only cause, these changing environmental conditions contribute to the continued spread of known diseases such as the tick-borne pathogens of **Lyme disease** and **viral meningitis,** and infections have increased significantly in Germany (Hemmer et al. 2018).

In addition, new infectious diseases are appearing in our country, the pathogens of which were completely new or, as "exotics", strongly linked to tropical or subtropical regions. Both humans and animals are affected by these. One example, which will not be discussed further later in this book, is **EHEC.** This disease had been known for some time, but occurred in a particularly severe form increasingly in humans in 2011 and affected northern Germany in particular. The pathogen was a pathogenic strain of the normally harmless bacterium *Escherichia coli*, which often occurs in the intestine. Although the pathogen was quickly detected in sick people, the source was initially unclear. It was not until some time later that an imported vegetable was identified as the source. The outbreak ended in 2011 (Karch et al. 2012). In this case, it was the ingestion of contaminated food. However, depending on the pathogen, there are also other **transmission routes** such as droplet infections, direct contact or transmission by insects (Fig. 1.1). The various infectious agents differ in many other aspects besides transmission.

P. U. B. Vogel, G. A. Schaub, *New Infectious Diseases in Germany and Europe*, essentials, https://doi.org/10.1007/978-3-658-41826-7_1

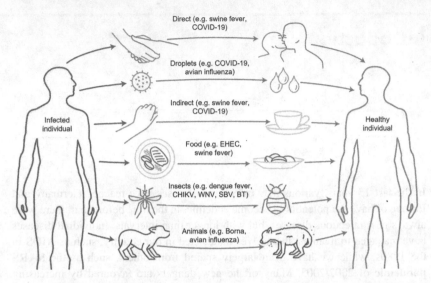

**Fig. 1.1** Types of transmission of infectious diseases. (Swine fever: African swine fever; EHEC: Enterohaemorrhagic *Escherichia coli;* CHIKV: Chikungunya virus; WNV: West Nile virus; SBV: Schmallenberg virus; BT: Bluetongue virus; Image source: Adobe Stock, file no.: 363871243, modified)

In this *Essential*, **infectious diseases** transmitted by direct and indirect contact are presented first. First, the biology, epidemiology as well as the economic importance of the dreaded **African swine fever** are described. The first occurrence in Germany had far-reaching consequences. For example, it led to an immediate import ban on German pork, among others by the largest customer China. The importance of so-called **zoonoses**, i.e. diseases that can be transmitted from animals to other animals or to humans, is then discussed. These include the frequently occurring **bird flu,** which has an enormous risk potential for the poultry industry and is brought to Europe, for example, by migratory birds, but also the very rare **Borna disease**. Both can also be transmitted to humans through direct contact and lead to a high case fatality rate in humans. In addition, avian influenza poses the risk of further development of the virus through genetic mixing with influenza A subtypes adapted to humans. Finally, the most significant zoonotic disease in the world at present is presented, **COVID-19.** This new infectious disease shows unlike the previous two examples, efficient human-to-human transmission after the probable jump from animals to humans. This disease spread very rapidly throughout the world and is unique in human history due to the duration of the pandemic and the measures taken to control it. The long-term consequences (**long covid**), which are unusual for respiratory diseases, and the question of origin are also highlighted.

Another large group of pathogens are the so-called **arboviruses.** These are viruses that are transmitted to animals or humans by insects, mostly mosquito species. The question of which of the mosquito-borne pathogens can also manifest themselves here in Germany depends crucially on the transmission potential of native mosquito species and the climatic conditions. For this purpose, we will deal with the biology and transmission potential of native mosquito species such as biting **midges** and the **common house mosquito** *(Culex pipiens).* Biting midges are important vectors of infectious diseases of animals. Changing climatic conditions also facilitate the establishment of exotic vectors. These include newly immigrated, so-called invasive mosquito species, such as the **Asian tiger mosquito** or the **Asian bush mosquito,** which can transmit certain tropical arboviruses. The spread of the Asian tiger mosquito is considered the greatest threat to the global spread of **dengue fever.** Another example are **giant ticks** (*Hyalomma* spp.), which can transmit diseases such as spotted fever and have been found more frequently in Germany in recent years (Chitimia-Dobler et al. 2019). However, we will not discuss ticks in more detail in this *Essential.*

Finally, Chaps. 5 and 6 present various examples of outbreaks and epidemics of arboviruses, respectively. The completely new **Schmallenberg virus** spread throughout Europe from 2011 onwards and caused damage primarily in cattle herds due to malformations in newborn calves. Another example is **bluetongue,** which had been known for some time but first appeared in Germany in 2006 and caused severe outbreaks until 2009. Intensive control measures brought this animal disease under control. Until 2018, Germany was considered free of bluetongue. However, this was followed by further outbreaks with varying frequency and extent (FLI 2021a).

The risk potential of new **human arboviruses** is also evidenced by local outbreaks, such as the **chikungunya virus** in Italy or the **dengue virus** in France, although immediate countermeasures were successful. **West Nile virus** also has epidemic potential. It is introduced by migratory birds, for example, and is already more common in Europe. In recent years, there have been isolated cases of lethal infections in humans, such as in Leipzig at the end of 2020. There are also initial indications that the West Nile virus is not only newly introduced every year, but already overwinters in local mosquito species in Europe.

In summary, this *Essential* presents some new **infectious diseases** of animals and humans in Germany and Europe. The biological and epidemiological basis of the pathogens and their vectors as well as the risk potential are described.

- African swine fever
- Zoonoses: avian influenza, Borna disease, COVID-19
- Local and new mosquitoes: biting midges, common house mosquito, Asian tiger mosquito and Asian bush mosquito
- Arboviruses of animals: Schmallenberg and bluetongue viruses
- Human arboviruses: chikungunya virus in Italy, dengue virus in France and West Nile virus in Europe

# African Swine Fever: Biology, Epidemiology and Economic Importance

2

**African swine fever** is a notifiable animal disease that is feared because of the damage it causes in pig herds, but also because of nationwide trade restrictions if it is detected. The causative agent is an enveloped DNA virus of the *Asfarviridae* family. The virus genome is encapsulated by proteins and additionally surrounded by a plasma membrane (= enveloped). The swine fever virus has a very large genome. Although the molecular biology of this virus is being intensively researched, many molecular aspects are still unknown, such as the cell receptor to which the virus binds for subsequent cell infection (Karger et al. 2019). The disease is severe in pigs, sometimes with non-specific symptoms but also severe bleeding (Fig. 2.1). There are different strains of the virus, some of which differ considerably in lethality. The **case fatality rate** can be up to 100% (Schulz et al. 2019).

This infectious disease is not a **zoonosis,** i.e. humans cannot contract it. Overall, the host range for this virus is quite limited, as it does not infect any other animal groups besides domestic and wild pigs (FLI 2021b). It is generally assumed that **African swine fever** spreads rapidly through the herd. However, the variation between different outbreaks is considerable and it has recently been questioned whether spread under field conditions is slower than generally assumed (Schulz et al. 2019). For example, in several outbreaks, only one animal or a few animals were infected at the time of detection. Further, infection was not spread to all animals kept in the same barn for more than a week (Chenais et al. 2019). In these cases, several factors seem to have been involved (virus isolate, virus titre, housing conditions, hygiene management). In contrast, other studies demonstrate infection of domestic pigs in direct contact with wild boar, but also when the animals are kept in different stables on the same farm, i.e. without direct contact (Guinat et al. 2016).

**Fig. 2.1** Bleeding that can occur in African swine fever. (Image source: Adobe Stock, File No.: 198402392)

**Transmission** in swine herds occurs directly from pig-to-pig orally or nasally via contact with infected animals, their excreta, or the remains of dead animals or contaminated surfaces (Cwynar et al. 2019; Mazur-Panasiuk et al. 2019). In addition to pig-to-pig transmission or via contaminated material, the virus can also be transmitted via **ticks**. In some regions of Africa, this is a typical mode of transmission, with leather ticks of the genus *Ornithodoros* being efficient vectors. Here, the virus circulates in so-called sylvatic cycles between ticks and warthogs and can then be transmitted back to pig herds (Bonnet et al. 2020). In contrast, tick species commonly found in Europe, such as *Ixodes ricinus* (common wood tick), do not appear to be suitable **vectors.** For this reason, this form of transmission is considered very unlikely in Europe (Guinat et al. 2016). The significance of another insect species, the fly ***Stomoxys calcitrans*** (stable fly), which is frequently found in animal stables, is still unclear. This fly feeds on blood and can transmit African swine fever virus mechanically for up to one day, i.e. through contamination of the body surface or mouthparts. In addition, the virus remains infectious in ingested blood in the stomach of the fly for about 2 days. Despite the unclear data situation, depending on the conditions, this fly is considered a risk factor to accelerate the spread of **African swine fever** (Bonnet et al. 2020).

Basically, **African swine fever** is not new to us. The disease first arrived in Europe in the late 1950s and spread throughout most of Europe over several decades (Cwynar et al. 2019). One example is an outbreak in Belgium in the 1980s that is believed to have been caused by imported contaminated meat. This outbreak affected about a dozen pig farms. As a preventive measure, over 30,000 animals on 60 farms were culled and the disease was declared conquered based on a subsequent screening of over 3000 farms (Biront et al. 1987). While the disease has been **endemic** in Africa since its discovery in the early twentieth century, it was considered eradicated in major European countries as of 1995. However, after the turn of the century, the virus reappeared for the first time in 2007 in Georgia, which is close to Europe. From there, it spread further westward, reaching Poland in 2014. Since then, the virus has been detected in thousands of wild pigs. Furthermore, there have been over 200 outbreaks of **African swine fever** in pig herds in Poland (Cwynar et al. 2019). In addition, the infectious disease was also detected in a feral pig in Belgium in 2018 (Sauter-Louis et al. 2020), although this was probably not caused via immigrant wild boars due to the long distance involved. In addition to Poland, Romania has also been severely affected. In this country, as in Poland, there have been outbreaks in pig farms in addition to detection in numerous wild boars. The **Friedrich-Löffler-Institut** keeps an overview of all detections in wild boar and pig herds for the affected European and bordering countries, including Germany (FLI 2021b).

In Poland, the **infectious disease** first spread eastwards. When the first cases were also found in the west of the country, the situation worsened, as the pig industry is mainly located here (Mazur-Panasiuk et al. 2019). Due to the common border with Poland, there were fears of entry into Germany as well. There were some preventive measures such as the erection of mobile border fences to prevent the crossing of wild boars. Epidemiologically, a finding a few kilometres to the left or right of the national border does not matter, but it is the **massive trade restrictions** that are to be feared. A positive detection, no matter how close to the border, will result in the whole country being classified as positive for **African swine fever**. In the end, despite preventive measures, the entry was not prevented and increased monitoring in the border region confirmed the first infected wild boar in Brandenburg and thus in Germany on 10 September 2020. The **gene sequence** of the first isolate was similar to that of Polish isolates. However, it is assumed that the virus was already introduced into Germany in July 2020 (Sauter-Louis et al. 2020). Only after the introduction was the construction of fixed border fences started in some regions (Fig. 2.2).

**Fig. 2.2** Fence on the German-Polish border to prevent the introduction of African swine fever by wild boars. (Image source: Adobe Stock, File No.: 397763172)

The event had immediate consequences. Germany is one of the main exporters of pork. Along with other countries, China, the largest buyer, issued an **import ban.** As a result, pork prices in Germany fell noticeably (tagesschau.de 2021). However, it did not remain a single case. In the weeks that followed, many more wild boar infections were confirmed (Sauter-Louis et al. 2020). Although Germany has effective **disease prevention** and control programs, African swine fever continued to spread in Brandenburg in the following months. Since January until the beginning of March 2021, almost 400 infected wild pigs have been confirmed, although no cases have yet been found in commercial pig herds (FLI 2021b).

A major problem is the combination of wild boars as carriers, the course of the disease and the stability of the virus. Wild boars can travel distances of 2–10 km each day, leading to rapid spread (Schulz et al. 2019). Infected animals excrete the virus with various excretions. The highest concentration of virus is found in the blood. However, the virus is also found in smaller amounts in saliva, urine and faeces (Guinat et al. 2016). In addition, the virus is enormously stable under a variety of conditions. In preserved ham, the virus remains **infectious** for 1 year. Similarly, carcasses of dead wild boars are a **source of infection** in which the virus remains infectious for a long time, even for years when frozen. In addition, animals continue to excrete the virus for several weeks after surviving the disease (Mazur-Panasiuk et al. 2019; Chenais et al. 2019). This high stability of the virus has also led, for example, to the emergence of the virus in new regions, with **contaminated meat products** playing a major role (Chenais et al. 2019). Experience with **African swine fever** shows that the disease in new regions is often not so easy to "stamp out", but is very difficult to control. For example, the outbreak in Belgium in the 1980s was caused by feeding contaminated meat to domestic pigs, and was thus confined to pig farms and thus more 'isolable'. In contrast, the situation is very difficult to control when the virus is circulating in the **wild pig population.** For this reason and the further possible renewed entries from neighbouring countries, this infectious disease will probably continue to pose a constant threat to Germany in the coming years, although active and passive monitoring have been greatly expanded.

# Zoonoses: Avian Influenza, Borna Disease and COVID-19

3

## 3.1 Avian Influenza

**Avian influenza** is a major viral disease of wild birds and poultry in livestock and is caused by **influenza A viruses**. Infections result in major economic losses in the poultry industry. Influenza A viruses are simplistically characterized by the composition of two surface proteins of the virus particle, hemagglutinin (H) and neuraminidase (N) (Fig. 3.1). There are 18 H types and 11 N types, which can occur in combination with each other, e.g. H5N1 or H7N9. Most of these types also occur in birds (Webster and Govorkova 2014). However, this composition cannot be used to infer how pathogenic a virus subtype is. There are low pathogenic (LPAI for low pathogenic avian influenza) and highly pathogenic (HPAI for highly pathogenic avian influenza) virus types. The highly pathogenic types cause the so-called avian influenza, also known as bird flu. At the beginning of this century, the most feared avian influenza agent was an influenza A strain of type **H5N1** (Webster and Govorkova 2014). The first major outbreak of this H5N1 subtype occurred in Hong Kong in 1997, followed by massive disease control measures including the culling of all flocks. After a few years, this subtype was present again, including in some European countries from the mid-2000s (Sellwood et al. 2007). This subtype has some zoonotic potential, with jump to humans occurring only in close contact (animal care or processing of infected animals or culling of infected herds). Nevertheless, this subtype is not well adapted to humans, and there has been no efficient human-to-human transmission. Approximately 850 human infections with this H5N1 type are known worldwide, with a **case-fatality rate** of over 50%, but none in Germany (RKI 2021).

    **Influenza A viruses** are constantly changing through various genetic mechanisms. One important mechanism is the so-called **reassortment** (Bouvier

© The Author(s), under exclusive license to Springer Fachmedien Wiesbaden GmbH, part of Springer Nature 2023
P. U. B. Vogel, G. A. Schaub, *New Infectious Diseases in Germany and Europe*, essentials, https://doi.org/10.1007/978-3-658-41826-7_3

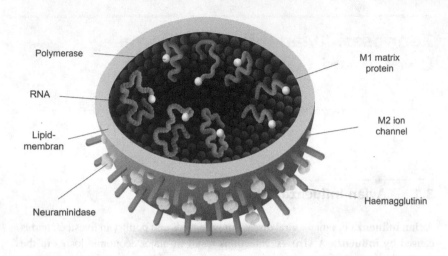

Polymerase

RNA

Lipid-
membran

Neuraminidase

M1 matrix
protein

M2 ion
channel

Haemagglutinin

**Fig. 3.1** Schematic cross-section showing the structure of an influenza A virus with important proteins. (Image source: Adobe Stock, file no.: 68969046, modified)

and Palese 2008). This mechanism is based on the principle that the **viral genome** of influenza viruses consists of different RNA strands, each of which codes for specific proteins (Fig. 3.1). If two different influenza viruses replicate in the same cells in an individual, the RNA strands of both types are formed. Here, due to the faulty packaging of new **virus particles**, an exchange of RNA strands can occur, the so-called reassortment. This is a typical mechanism by which influenza subtypes with new properties can arise spontaneously. The **H5N1 subtype** has also emerged through reassortment (Webster and Govorkova 2014). However, the importance of this subtype has declined sharply in Europe in recent years. However, there are now several new influenza A subtypes that can cause **avian** influenza. These are mostly based on **H5** in combination with various N types. In the last 15 years, several of these new subtypes have caused outbreaks (Verhagen et al. 2021).

These viruses are often carried by **migratory birds**, which, depending on the season, travel long distances to reach their breeding grounds or habitats. This behaviour makes them ideal distributors for viruses, including **influenza viruses**. During resting or return, they may transmit the viruses to wild birds, which in turn may sporadically transmit them to livestock (Globig et al. 2018). In addition, there are other ways of introduction, such as through illegal animal trade, animal transport, etc.

**Fig. 3.2** Laying hens in farm animal husbandry. (Image source: Adobe Stock, file no.: 199602338)

In 2014, a new strain was discovered that arose, among other things, through reassortment from the **H5N1 subtype**. Following the spread of this **H5N8 type** in China and elsewhere, the first outbreaks occurred in Europe, including Germany, at the end of 2014. The exact route of entry is unclear, but in addition to the above-mentioned causes, an unusually cold weather in Russia in 2014 may have pushed many birds westwards, introducing the virus (Adlhoch et al. 2014). The particular danger of the H5 types is that they circulate in poultry as low pathogenic variants, but can quickly develop into highly pathogenic types (Verhagen et al. 2021).

The governmental measures in case of detection of **avian influenza** are substantial and include the establishment of a **restriction zone** and a limitation of mobility within this zone and to other areas. Furthermore, the affected poultry flocks will be culled prophylactically. In poultry, the risk of rapid spread is particularly high due to the high density of the flock (Fig. 3.2). The draconian measures are necessary because avian influenza would otherwise spread rapidly and thus pose a threat to livestock farming as a whole, but also to humans. So far, there have only been a few cases of avian influenza spreading to humans. However, a high proportion of these have been fatal. Influenza A subtypes that are well adapted to humans and endemic in many places are limited to H1N1, H2N2 and H3N2 (Uyeki and Peiris 2019).

In the past, outbreaks of highly pathogenic **influenza A strains** tended to be confined to poultry flocks and also geographically, without wild birds playing a significant role. This situation has changed in recent years and decades. Nowadays, contact between poultry flocks and wild birds already poses a greater risk (Globig et al. 2018; Verhagen et al. 2021).

The most significant **avian influenza epidemic** in the history of Germany occurred in 2016/2017. There were over 1000 detections of highly pathogenic influenza A variants, mainly H5N8 distributed across Germany. During this period, there were over 100 outbreaks in poultry flocks in farm animal husbandry or in zoos. The pathogen spread from Russia via Hungary and Poland to Germany (Globig et al. 2018). This epidemic was most likely not caused by a single entry event, but by repeated entries. This epidemic also did not involve a homogeneous virus subtype, but several influenza A types created by **reassortment** (Pohlmann et al. 2018).

Even though the situation in Europe has been somewhat calmer in recent years, the situation remains tense. In 2020, there were 290 outbreaks of highly pathogenic avian influenza in Europe in a 3-month interval (Feb.-May). The outbreaks occurred predominantly in Hungary; however, there were isolated cases in Germany. In most cases, the **H5N8** subtype was identified (European Food Safety Authority et al. 2020). In the early years, the **H5N8 subtypes** did not pose any zoonotic potential, meaning that these strains did not spread to humans (Globig et al. 2018). This is likely due to the fact that these virus subtypes are not a good match for humans in terms of host cell receptor specificity (Adlhoch et al. 2014). However, some cases of human disease starting from an H5N8 subtype were found in Russia in late 2020 (RKI 2021). Among the newer circulating H5 types, besides **H5N1,** there is only **H5N6,** which caused severe human disease in isolated cases in China (Uyeki and Peiris 2019). This type has been found very rarely in birds in Germany since the beginning of 2018 (RKI 2021).

A current example of **avian influenza** is Schleswig-Holstein, which is regarded as a particular hotspot in Germany because it is a particularly frequent overflight area or port of call for migratory birds. In the district of Plön, almost 80,000 laying hens already had to be culled in March 2021 because avian **influenza** was detected in the flock, which was caused by an **influenza A virus of** the type **H5N8** (NDR 2021). Furthermore, there are new influenza A subtypes that can also be dangerous for humans and have a high case fatality rate, such as the **H7N9** found in China (RKI 2021). These variants have not played a role in Germany to date.

## 3.2     Borna Disease: Very Rare and Always Fatal?

**Borna disease** is a very rare zoonotic infectious disease that occurs worldwide, including Germany and other European countries. The disease is not restricted to a specific species. In addition to horses and humans, sheep and cattle, for example, can also be affected (Ludwig and Bode 2000). The disease is caused by **Borna virus**, an enveloped RNA virus. The human disease manifests itself, among other things, in meningitis and can be fatal, but mainly affects people whose immune system is weakened. The symptoms of the disease are not caused by the multiplication of the virus, but by a specific immune response that is too strong (Richt et al. 1997).

Even though, curiously, this disease has only been published more frequently in scientific journals since the mid-1990s, **Borna disease** is not all that new. The disease was first noticed in 1885, when a large number of horses died in the town of Borna in Saxony. At that time, however, the pathogen was unknown (Richt et al. 1997). Borna disease is not the only viral infectious disease in horses that causes neurological damage. It is one of a number of rare or even rarer infectious diseases affecting the central nervous system, such as rabies, equine herpes and **West Nile fever**, discussed in Sect. 6.3 (Lecollinet et al. 2019). Increased attention has also been accompanied by improvements in the diagnostic field. In Austria, too, several cases have been reported in recent years in which horses have demonstrably fallen victim to the disease (Weissenböck et al. 2017).

**Borna disease** has been classified as a predominantly fatal zoonosis. However, a large number of seropositive individuals are also said to be present (Rubbenstroth et al. 2020). Because of this, there have been suggestions that many people have already come into contact with the pathogen, thus the high assumed case fatality is overestimated due to the focus on clinical severe cases. However, a broad-based study in southern Germany revealed a low seroprevalence (Tappe et al. 2019).

Epidemiologists and physicians are interested, as so often in such cases, in the source and transmission. The question here is how the virus is transmitted to horses or humans. A study in Switzerland identified shrews as a **reservoir** (Hilbe et al. 2006). In biology, a reservoir is defined as a species of animal (or human) that usually carries the pathogen asymptomatically and can pass it on when it comes into contact with other species. The exact **mechanism of transmission** is not yet fully understood. However, there are data suggesting infection via the nasal mucosa (e.g. after inhalation), followed by **viral replications** in neurons of the olfactory system and migration to the central nervous system (Kupke and Becker 2019).

## 3.3    COVID-19

In December 2019, the metropolis of Wuhan, China, with millions of inhabitants, had the first patients with a new unusual lung disease (Xu et al. 2020). In early January, a new **coronavirus** was found to be the cause, **SARS-CoV-2**, and the virus was spreading at a rapid rate around the world. At the time of writing (as of 18 May 2021), over 163 million infections had been laboratory-confirmed worldwide, with approximately 3.4 million deaths (CSSE 2021).

A total of seven **human coronaviruses** are known. Four of these coronaviruses (designated 229E, NL63, HKU1 and OC63) occur worldwide and cause typical colds during the cold season, which can, however, be dangerous for immunocompromised individuals. These coronaviruses then cause approximately 15% of all colds during this season (Kahn and McIntosh 2005; Greenberg 2016). The first human coronaviruses were discovered in the 1960s (Kahn and McIntosh 2005). However, molecular biology analyses indicate that some coronaviruses were transmitted to humans several hundred years ago and have been circulating in our population ever since. The last of these 4 types is thought to have jumped to humans around the end of the nineteenth century (Graham et al. 2013). However, two of these types were only discovered after the **SARS pandemic** due to improved coronavirus diagnostics.

In addition, there are three other **coronaviruses** that were only transmitted from animals to humans or identified after the turn of the century. The diseases caused by them had or have a special medical significance because of the high mortality rate. **Severe acute respiratory syndrome (SARS)** first appeared in China in November 2002. It resulted in 8096 cases with 774 deaths (WHO 2004). At that time, coronaviruses were unknown as severe human pathogens, and it took some time before a coronavirus was identified as the cause of SARS (Drosten et al. 2003). The next severe infectious disease caused by coronaviruses was **Middle East Respiratory Syndrome (MERS)**, first described in 2012 (Zaki et al. 2012). MERS is a constant threat as the virus circulates massively in dromedaries (one-humped camels) and is always sporadically transmitted to humans. In total, there are now over 2500 known human cases of MERS (WHO 2020). The case fatality rate is even higher than SARS, currently at 34.3% (WHO 2020). However, these two coronaviruses were or are not comparable to **SARS-CoV-2**, as human-to-human transmission was much less efficient. For this reason, the health consequences of these diseases were or are limited, in contrast to **COVID-19.**

A major aspect that helped **COVID-19** to overcome efficient **disease control** measures (identification of virus carriers, isolation and contact tracing) was the broad spectrum of subclinical and clinical courses. On the one hand, this disease sometimes leads to severe to fatal courses in risk groups such as the elderly or people with pre-existing conditions. On the other hand, there is a high proportion of people in whom the disease runs **subclinically,** i.e. unrecognised or like a common cold. In the beginning, a spreading epidemic can only be contained efficiently with the classical means of disease control if infected persons are detected. Subclinical infections run under the radar and lead neither to a report to the public health department nor to laboratory diagnostic tests and thus not to isolation. This can create undetected chains of infection. This is a major difference from other infectious diseases, where classical disease control works well. **SARS** and **Ebola patients** were characterized by a much more severe and homogeneous clinical symptomatology, although subclinical courses could or do occur (Wilder-Smith et al. 2005; Kuhn and Bavari 2017). This facilitated the detection and early isolation of infected individuals. In addition, these viruses were or are less efficiently transmitted.

Another peculiar phenomenon of **COVID-19** is the unusually high rate of protracted complications after recovery. The late effects are summarized as **long covid** or **post-COVID syndrome.** These late effects include shortness of breath, fatigue, chest or joint pain, or depression. Some COVID-19 patients from Wuhan experienced long-term sequelae for more than 6 months (The Lancet 2020a). Monitoring of over 100 patients in Israel who had undergone mild COVID-19 disease showed a similar picture. 6 months after recovery, just under half of the individuals reported secondary symptoms ranging from fatigue to impaired sense of smell and touch to difficulty breathing (Klein et al. 2021). This is of particular concern as even mild courses are not exempt from late effects. In addition to the aforementioned long-term symptoms, it is the neurological damage caused by COVID-19 that is the object of intensive research. It is currently being investigated whether COVID-19 can also trigger, for example, **autoimmune diseases** of the central nervous system or neurodegenerative diseases such as **Alzheimer's disease** (Wang et al. 2020). From today's perspective, the rather unusual clustering of long-term sequelae raises the question of whether COVID-19 is really so unique or whether this has also been observed in other serious infectious diseases. Interestingly, historical data suggest that other epidemics and pandemics may have been associated with severe neurological late effects, such as **Spanish flu** or **diphtheria** (Stefano 2021). Ultimately, it will take time for the full picture of Long Covid to emerge through the follow-up of COVID-19 patients and intensive research into its molecular basis and pathogenesis.

At present, it looks as if the **COVID-19 pandemic,** which has so far been kept in check by almost worldwide lockdown measures, can be brought under control by the worldwide use of vaccines and massive testing. There are also several different **vaccines** already licensed and successfully deployed in the EU (Vogel 2021). Despite this tremendous success, the damage is already significant, not to mention the fact that COVID-19 will rage in certain regions for some time before everyone is vaccinated. The full extent and aftermath of this health, social, economic, and political damage will likely not become apparent for years, from perhaps as yet unknown secondary effects of COVID-19 disease to psychological effects of contact restrictions to protracted bankruptcies or dangerous shifts in political power.

The cause of the **COVID-19 pandemic** is particularly important in terms of whether and how we can protect ourselves from future coronavirus pandemics. Early on, transmission from bats to humans was hypothesized by the Chinese government. The viral genome of **SARS-CoV-2** had 96.2% sequence identity to a coronavirus from bats in the region (Guo et al. 2020; Yuen et al. 2020). The conjecture is obvious and is also currently supported by a WHO commission, although it is likely that another host after the bat led to the infection of humans (Maxmen 2021). In the aftermath of the **SARS pandemic,** the central role of bats in the emergence of new human pathogenic coronaviruses was recognized. Based on sequence analyses, all known human coronaviruses are thought to have originated in either bats or rodents, presumably with some direct transmission to humans or with intermediate other animals as so-called intermediate hosts (Corman et al. 2018). However, significant divergence from the sequence of SARS-CoV-2 is present in viruses found in bats (Ye et al. 2020). A broader genetic analysis suggests that SARS-CoV-2 originated in a different area and reached the animal market in Wuhan through transport (Harapan et al. 2020). Certain pangolins such as scaly anteaters have been implicated as intermediate vectors (Guo et al. 2020), although the coronavirus sequence identity of these animals is even lower than that of bats (Ye et al. 2020). In addition, in the early stages of the outbreak, individuals who had no connections to the animal market also tested positive (Peeri et al. 2020). Overall, these studies show that elucidating **spillover events** is often akin to searching "for a needle in a haystack."

In addition to the **biological cause,** political and cultural factors must also be taken into account. In the case of **SARS,** the Chinese government was heavily criticized for its inadequate and non-transparent communication. After the initial notification, WHO did not receive further information from the Chinese government until months later in March 2003 (Peeri et al. 2020). By this time, SARS had spread to many countries (Gillim-Ross and Subbarao 2006). Even though the sequence of **SARS-CoV-2** was provided very early in January, the detention of health workers warning of this new disease still occurred in China (The Lancet 2020b). The risk of emergence of new **zoonotic diseases** must also take into account general cultural

**Fig. 3.3** Live market in China. (Image source: Adobe Stock, file no.: 318222444)

practices. Trade in live and wild animals in markets has a long tradition in China and elsewhere (Peeri et al. 2020; Fig. 3.3). In this context, the risk of zoonotic **spillover events** is not limited to **coronaviruses.** Transmission of avian influenza pathogens such as H5N1 and H7N9 to humans, often with fatal outcomes, is also known from these markets (Morens et al. 2020). The proposal to close these markets does not have much chance of success, as this tradition is culturally ingrained and the income of many families depends on it (Peeri et al. 2020). Certain cultural habits also hindered, for example, the containment of the Ebola epidemic of 2013–2016 (Vogel and Schaub 2021). However, perhaps the ongoing encroachment of the livestock industry into wildlife habitats could be halted through international agreements and funding support. This is certainly a factor that would reduce the "clock rate" of new spillover events in the future. In addition, intensive regular serum monitoring of workers in these markets or farms could help to detect new potentially pandemic virus variants at an early stage.

Fig. 3.2 ...

# Mosquitoes: Species, Biology and Epidemiology

**4**

## 4.1 Native Species: Biting Midges and Common House Mosquitoes

In Europe there are thousands of species of different flying insects of the order of the Diptera, all possessing two wings. The **mosquitoes** represent a suborder (Nematocera) with over 40 families. The term family originates from biological systematics and refers to a group with the same phylogenetic roots, but whose members may differ significantly phenotypically. In systematics, animals are assigned to the levels order, family, genus and species. For example, the **common house mosquito** (*Culex pipiens*), which is particularly widespread in Germany, belongs to the same family (Culicidae) as the species *Anopheles gambiae,* one of the most efficient vectors of malaria. However, they belong to different genera (*Culex* vs. *Anopheles*) and also have differences in their biology (appearance, reproduction, habitats, host preference, etc.). Of the more than 40 families of mosquitoes, two receive special consideration in this Essential because they are involved in the transmission of the infectious diseases presented here. These are the **biting midges** (family Ceratopogonidae) and the **mosquitoes** (family Culicidae).

A distinction is made between two different mechanisms by which transmission by mosquitoes can occur. In the case of mechanical transmission, for example, the surface of the insect is contaminated with the pathogen and transmitted to the next host by contact. This is suspected, for example, in the case of **African swine fever** (Chap. 2). In the case of **arboviruses,** this mechanism is insignificant. Here, the pathogens are actively transmitted by mosquitoes. The ability to actively transmit a virus or other organisms such as parasites is called **vector competence.** The question of whether a particular mosquito species can transmit a virus depends on many factors. In the simplest example, the virus must be able to establish and

© The Author(s), under exclusive license to Springer Fachmedien Wiesbaden GmbH, part of Springer Nature 2023
P. U. B. Vogel, G. A. Schaub, *New Infectious Diseases in Germany and Europe*, essentials, https://doi.org/10.1007/978-3-658-41826-7_4

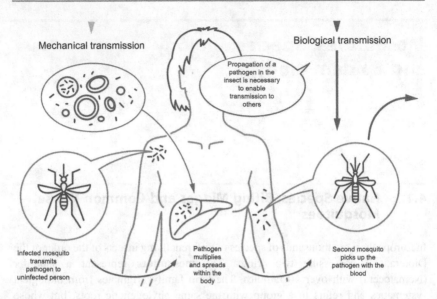

**Fig. 4.1** Transmission mechanisms of infectious diseases by insect vectors. (Image source: Adobe Stock, file no.: 353858026, modified)

replicate in the mosquito, i.e. actively enter and replicate in cells of the mosquito's gut. Furthermore, these viruses must spread in the mosquito's body and enter the **salivary glands.** Only then can a mosquito inject the viruses into the wound with its saliva during the next blood meal. After such an infection, the virus must then be able to multiply strongly in the new host and be present in the blood in order to be picked up by the next mosquito. Only when these two conditions are met can efficient mosquito-to-mosquito transmission cycles occur (Fig. 4.1). However, there are other factors such as ambient temperature that affect transmission ability (Vogels et al. 2016). Knowledge about the **vector competence** of certain mosquito species for certain viruses is still incomplete and a transmission ability is often assumed without really being proven (Sick et al. 2019).

**Biting midges** are a family (Ceratopogonidae) that play a special role in the transmission of infectious diseases, especially in animals. These Diptera are also widespread in Germany and Europe. During a monitoring of biting midges in Europe during the **bluetongue disease** outbreak in 2006–2009, biting midges were found in all climatic zones of Switzerland (Kaufmann et al. 2009).

**Biting midges** are very small mosquitoes with a size of 1–3 mm (Ayllón et al. 2014). Carried by the wind, they can travel quite long distances of 0.9–1.5 km per day (Endalew et al. 2019). Biting midges develop similarly to mosquitoes from egg through larvae (4 larval stages) and pupa to adult insect, with temperature-dependent development taking several weeks. Biting midges are quite short-lived, with a life span of 10–20 days (Sick et al. 2019). The biology of biting midges is less studied than, for example, that of *Anopheles* mosquitoes, which transmit the malaria pathogens, among other diseases. In recent years, however, scientists have gained a somewhat clearer picture regarding the pattern of activity and preference for certain hosts. Certain species of biting midges are found in greater numbers from mid-March onwards, depending on temperature, and have a much higher preference for cattle than for horses, for example, so are found much more frequently in **cattle herds** (Kameke et al. 2017). Biting midges are predominantly active at dawn and sunset, with little activity during the day (Ayllón et al. 2014). At these peaks of activity, they often occur in large swarms (Fig. 4.2).

The question of why **biting midges** prefer to settle in certain regions is still unclear. Manure storage practices are thought to have an influence on the occurrence of biting midges (Werner et al. 2020). Although biting midges are also known

**Fig. 4.2** Swarm of midges infesting highland cows on pasture. (Image source: Adobe Stock, file no.: 167907070)

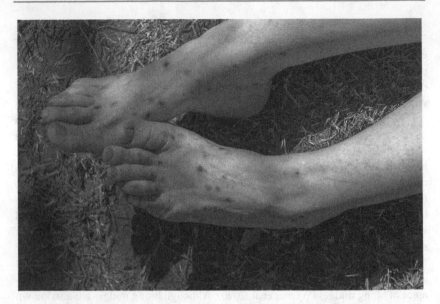

**Fig. 4.3**  Numerous painful gnat bites. (Image source: Adobe Stock, file no.: 262660432)

to transmit other veterinary diseases, humans need not fear them. However, they can cause painful sores (Fig. 4.3). In the family of biting midges, it are mainly the members of the genus ***Culicoides*** that are considered to be of particular importance in the transmission of **arboviruses.** However, the European species cannot be studied in the laboratory because they do not reproduce under laboratory conditions (Sick et al. 2019).

The **common house mosquito** (*Culex pipiens*) belongs to the family Culicidae and is one of the most common mosquitoes in Germany, depending on the area. The species occurs in two biotypes (*C. pipiens* biotype *pipiens* and *molestus*, respectively). The two externally hardly distinguishable biotypes differ in many aspects, e.g. breeding sites, **host preference** (*pipiens* prefers birds, *molestus* among others humans) and the way of hibernating, i.e. continuing to be active or going into a resting phase (Koenraadt et al. 2019). In addition, the species *Culex torrentium* is also widely distributed in Germany (Becker et al. 2012). These species are also frequently found in dwellings. These mosquitoes are relatively small, measuring about 5 mm, and are characterized by particularly long legs (Fig. 4.4). As with almost all mosquito species, only females feed predominantly on blood. For this purpose, they have so-called piercing-sucking mouthparts, i.e. a proboscis whose

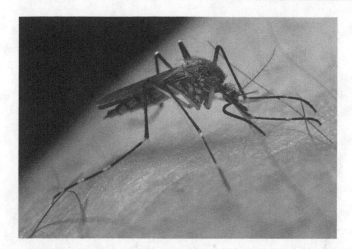

**Fig. 4.4** Common mosquito sucking blood. (Image source: Adobe Stock, file no.: 253760504)

inner parts pierce the skin to suck up blood. The females need the **blood meal** for reproduction. They then lay several hundred eggs on or near water surfaces. The entire development (egg, larvae, pupa) to the adult mosquito depends on the temperature and takes on average about 3 weeks.

Despite the aforementioned limited knowledge of vector competence, there have been more and more scientific studies on this in recent years. For example, two *Culex* species, *Culex pipiens* and *Culex torrentium*, are susceptible to **West Nile virus** infections in Germany (see Sect. 6.3) and are presumably also capable of transmitting it (Leggewie et al. 2016; Jansen et al. 2019). The same species are also capable of transmitting **Usutu virus** in Germany, as the virus was present in saliva at 25 °C 2–3 weeks after experimental infection (Holicki et al. 2020). Usutu virus is another **arbovirus** not detailed in this *Essential* that may receive greater attention in the future. *Culex tarsalis* also allows efficient propagation of West Nile virus (Dunphy et al. 2019). Similarly, feeding experiments with *Culex quinquefasciatus* suggest a likely transmission capability for West Nile virus (McMillan et al. 2019).

## 4.2  **Invasive Species: Asian Tiger Mosquito**

The **Asian tiger mosquito** (*Aedes albopictus*) also belongs to the family of mosquitoes (Culicidae). At up to 10 mm, this mosquito is somewhat larger than the **common house mosquito** and is characterised by its black and white ringing

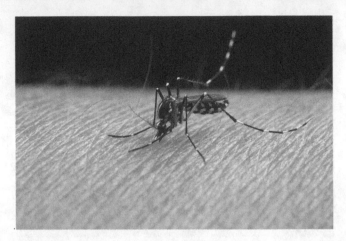

**Fig. 4.5** Asian tiger mosquito *(Aedes albopictus)* sucking blood. (Image source: Adobe Stock, file no.: 308686947)

(Fig. 4.5). Therefore, it is often confused with the native ringed mosquito. The tiger mosquito was formerly native to subtropical and tropical regions and originated in Southeast Asia (Johnson et al. 2018). However, it has been spreading increasingly in Europe in recent decades, often via the transport of commercial goods such as used car tyres (used for "whispering asphalt"), but also via vehicle traffic. The reason for their establishment in new areas is their high adaptability, among other things due to the eggs being resistant to desiccation. The species is now considered endemic in about 20 European countries (Bundesumweltamt 2019). In some regions, it has already overtaken the common house mosquito. For example, the Asian tiger mosquito was the dominant mosquito species (almost 90%) in a monitoring at different trapping sites in Turkey near the Black Sea (Akiner et al. 2019).

The Asian tiger mosquito was first found in Germany in 2007. In addition to the monitoring of certain research groups and institutions, since 2012 there has been the so-called **Mosquito Atlas Project,** which is led by the **Leibniz Centre for Agricultural Landscape Research** (ZALF) **and the Friedrich-Loeffler-Institute** (FLI). In this project, anyone can send in a mosquito caught at home or outdoors. The mosquito is then morphologically identified and its occurrence mapped. As part of this project, the **Asian tiger mosquito** was discovered in new locations in Germany, although this species is still a minority overall (Walther and Kampen 2017). Nevertheless, the risk of rapid spread should not be underestimated. For example, the Asian tiger mosquito was first detected in Spain in 2004 in a town in the northeast of the country. Within 10 years, it has spread all along the east coast, but also further inland (Johnson et al. 2018). The spread of this mosquito species is particularly dangerous because it is a "broad-spectrum vector" capable of transmitting a large number of **arboviruses** (including **dengue virus, chikungunya virus, West Nile virus**) under both laboratory and field conditions, depending on the region, whereas the **common house mosquito** is not capable of transmitting dengue or chikungunya virus (Martinet et al. 2019). The Asian tiger mosquito is often considered the greatest threat to the global spread of dengue fever due to its dispersal, which was previously restricted to the tropics by the main vector, the **yellow fever mosquito** (*Aedes aegypti*).

## 4.3   Invasive Species: Asian Bush Mosquito

The **Asian bush mosquito** *Aedes japonicus* (subspecies *japonicus*) from the family of mosquitoes (Culicidae) also originates from the Asian region. It is rather dark brown and thus differs in appearance from the **Asian tiger mosquito.** This also invasive species was first discovered in Europe in 2000 during a monitoring of the mosquito population in France. The mosquito larvae were found in a used car tyre. In 2008, it was also found in Germany (Koban et al. 2019). There are some geographically restricted populations from the north to the south of Germany, most of which were also discovered by the **Mosquito Atlas Project** (Walther and Kampen 2017). These geographically separated populations showed a different dispersal pattern in the following years. While the population in the north is slightly declining, the population in the west is expanding its range significantly and covers the whole of Baden-Württemberg as well as parts of North Rhine-Westphalia, Rhineland-Palatinate and Hesse (Koban et al. 2019). Therefore, the Asian bush mosquito can no longer be considered a small minority and it can be estimated that it will spread to large parts of Germany within a few years. The Asian bush mosquito is also very adaptable, as it has no pronounced preference for specific breeding sites and lays its eggs in tree holes, but also in vases or buckets that provide enough water, preferably in the transitional area between deciduous forests and inhabited areas (Kampen et al. 2016; Früh et al. 2020). It is active longer in autumn than the common mosquito (Früh et al. 2020). However, there are also limits to this. Hibernating mosquito eggs require temperatures $\geq 5$ °C and die at lower temperatures (Reuss et al. 2018).

The Asian bush mosquito, just like the Asian tiger mosquito, is dangerous in terms of future establishment of tropical and subtropical **arboviruses.** Possible transmission of **West Nile**, **dengue** and **chikungunya viruses** has also been confirmed for this species under laboratory and field conditions (Martinet et al. 2019).

# Arboviruses of Animals: Schmallenberg and Bluetongue Viruses

**5**

## 5.1 Schmallenberg Virus: Spread of a New Pathogen in Europe

The **Schmallenberg virus** presented in the following is a prime example of how a completely new virus can initially manifest itself unnoticed in areas such as Europe and then spread playfully throughout Europe. In late summer 2011, a few cases were observed in cattle herds in North Rhine-Westphalia, with fever and a reduction in milk production as clinical symptoms. A veterinarian from Baden-Württemberg investigated the cases and reported this to the **Friedrich-Loeffler Institute** (FLI). In addition, there were cases of miscarriages and malformations in newborn calves in various herds in Germany (Kupferschmidt 2012). The FLI initially tested for known pathogens that would have been consistent with the clinical signs, but without success. By applying the so-called **"deep sequencing"** method, in which all nucleic acids present in the sample are decoded, sequences of a novel virus were found that showed similarities to so-called **orthobunyaviruses.** This group includes viruses such as Akabane virus, which is found in Japan and Australia and is known to cause similar clinical symptoms in cattle, sheep and goats. However, orthobunyaviruses had never been detected in Europe at that time (Tarlinton et al. 2012).

The virus was named after the town of **Schmallenberg** in North Rhine-Westphalia because the virus was first detected in samples from this town. The **Schmallenberg virus** is an enveloped **RNA virus.** The virus genome contains 3 RNA strands that code for a total of 6 proteins. In addition to the so-called structural proteins that are part of the virus particle, e.g. two surface proteins, there are three so-called non-structural proteins that are only produced during cell infection. One is the polymerase, which duplicates the viral genome so that many progeny can be formed in each infected cell. The other two non-structural proteins are mainly responsible for the **virulence** of the virus by interfering with signaling from infected cells, thereby interfering with the immune response to the virus (Endalew et al. 2019). In addition, one of the two viral surface proteins has highly variable domains that may also be involved in subverting the immune response (Collins et al. 2019).

Interestingly, it is completely unclear where the **Schmallenberg virus** comes from. The virus was never described before 2011. Related viruses are common in Asia and Japan, but do not have enough similarity to be direct ancestors of Schmallenberg virus (Endalew et al. 2019). Curiously, it occurred in the same region where **bluetongue** was first detected in Germany a few years earlier in 2006 (Claine et al. 2015; para. 5.2). It is possible that the virus emerged in regions where medical monitoring of animal herds was less pronounced and other animal pathogens caused greater damage, followed by an accidental entry into Germany. In such a scenario, a low pathogenic virus would have the possibility to remain undetected. In addition, it is also possible that the virus previously mutated from a non-pathogenic type to a type with higher virulence. On the other hand, arboviruses, like Schmallenberg virus, are genetically quite stable, so they are not actually mutation masters (Collins et al. 2019).

The disease caused by the virus is characterized by a short viremia phase of 1–6 days. During this time, the amount of virus in the blood is very high. The infection leads to fever, diarrhoea and reduced milk yield. However, the virus has the most severe effects on the offspring. In the fetus, the virus can multiply and cause various damages to the central nervous system, resulting in **miscarriages** and also **malformations** of the newborn calves and lambs (LGL 2021). In new infectious diseases, the transmission of the pathogen is an important issue. The virus has been detected by molecular biology in various excretions of animals such as faeces and urine. However, there is no evidence yet that transmission is direct from animal-to-animal (Endalew et al. 2019).

Shortly after the discovery of the virus in Germany, certain mosquitoes, the **biting midges,** came into sharper focus (Tarlinton et al. 2012). This suspicion was consistent, as many other orthobunyaviruses are also transmitted via biting midges (*Culicoides* spp.). The suspicion was increasingly confirmed as the virus was also commonly present in biting midges in many regions where the disease occurred (LGL 2021). However, transmission has not yet been experimentally proven.

The **Schmallenberg virus** spread "as fast as the wind" in Germany and also in neighbouring countries. Thus, the virus was also detected in the Netherlands as early as late summer 2011 (Tarlinton et al. 2012). It is assumed that the virus was already circulating in Germany in the spring or summer of 2011, except that the severe consequences such as miscarriages or malformations were only detected after the calves were born (LGL 2021). Thus, the virus had enough time to spread unnoticed. The **distribution area** of the Schmallenberg virus extended to large parts of Europe by the end of 2012 and continued to expand subsequently (Collins et al. 2019). The term "as fast as the wind" also applies in the truest sense of the word, as biting midges can travel long distances on the wind and thus reach new areas (para. 4.1).

For new viruses introduced into a new area by a single or a few events, certain conditions must be present for these viruses to become endemic, i.e. indigenous. Transmission by **competent vectors** such as **biting midges** is one requirement, but the usual season of biting midges is limited from spring to autumn. Therefore, the pathogen must also overwinter in order not to disappear again after just one season. This question has not yet been fully clarified in the case of **Schmallenberg virus** and biting midges, and there are various hypotheses. Some studies have found a minimally small active population of biting midges in cattle barns even in winter, where the virus could persist. Other studies found biting midges-free periods in livestock barns in winter through mid-March (Kameke et al. 2017). There is also evidence that the virus is transmitted by female biting midges so-called vertically to the eggs, which could allow the virus to survive the winter. However, this hypothesis is not yet widely accepted.

**Vaccines** are now available for the prevention of this infectious disease in cattle and sheep. The first approved vaccines were based on the inactivated pathogen. In the meantime, other candidates such as genetically modified live or DNA vaccines are on the way. However, the **vaccination rate** is insufficient. The first approvals were achieved after there was already a high **seroprevalence** due to natural infections. For this reason, the willingness to vaccinate their own herds was weak (Wernike and Beer 2020). How long **herd immunity** lasts will be important here. Based on the antibody response of recovered animals, an immunity of about 3–6 years is estimated. This estimate is also consistent with the epidemic occurrence of the related Akabane viruses in Japan at intervals of approximately 5 years (Endalew et al. 2019). In contrast, an **epidemic pattern** at 2–3 years intervals has been preliminarily observed for Schmallenburg virus. For this reason, the lack of **willingness to vaccinate** on the part of farmers is dangerous (Wernike and Beer 2020).

## 5.2   Bluetongue Virus

**Bluetongue disease** is also a notifiable animal disease caused by a virus of the family *Reoviridae*. The virus is an RNA virus of which 24 serotypes are currently known. The term serotype describes strains of virus whose surface antigens (primarily the so-called VP2 protein) differ in such a way that they can be classified by antibodies (Wilson and Mellor 2009). Bluetongue disease primarily affects cattle, goats and sheep, but is harmless to humans. Transmission is not animal-to-animal, but similar to **Schmallenberg virus** via **biting midges**. Although cattle dominate the infection figures, the damage in sheep flocks is more severe, with case mortality rates of up to almost 40% (Conraths et al. 2009).

Although the disease is subclinical in the majority of infected animals, there are also symptoms such as fever, lameness, oedema and even death. The eponymous symptom, blue tongue, occurs only occasionally in severe courses (Wilson and Mellor 2009).

Before 1998, **bluetongue disease** was restricted to certain regions in Africa, the Middle East, Asia, Australia and South and North America (Wilson and Mellor 2009). During this time, isolated outbreaks occurred in Europe. Since then, the disease has become increasingly naturalized and endemic in countries in southern Europe (Hagenaars et al. 2021). The disease first appeared in Germany and neighbouring countries in 2006. Here, it was serotype 8 that was dominant (Mehlhorn et al. 2009). In contrast to the **Schmallenberg virus**, the disease was known beforehand and there were diagnostic methods for detection and already established monitoring programmes.

The 2006 **outbreak** mainly affected North Rhine-Westphalia, but also spread to other federal states, affecting a total of almost 900 cattle and sheep herds. After a dormant period in winter, infections were detected again from June 2007 onwards. The virus spread throughout Germany and there were over 20,000 outbreaks. At the beginning of 2008, the disease was contained with massive **vaccination campaigns.** As a result, there were only just over 1000 infections in the course of the year (Conraths et al. 2009). It was a coincidence that the numbers were so low in 2008. Vaccines were quickly approved for prevention. However, the vaccination campaigns started in May, which was actually too late to prevent a major spread. Fortunately, because of the long winter, biting midges development started 2–3 months late (Mehlhorn et al. 2009). This gave the vaccination campaigns a time advantage that averted worse.

For many of the European outbreaks, animal transports of infected herds and the already described passive spread of the **biting midges** by the wind have been assumed. However, the cause of the introduction of **bluetongue disease** into Germany is unknown (Wilson and Mellor 2009). In 2012, Germany regained bluetongue disease-free status, however, the disease re-emerged from 2018 onwards (Hagenaars et al. 2021). However, the infection numbers are in different dimensions compared to the period 2006–2009. Numbers are low at 1 (2018), 59 (2019), 2 (2020) and so far 1 in 2021 (FLI 2021a), probably due to available vaccines and improvements in monitoring and containment measures. Nevertheless, the **economic costs** are immense, partly due to the damage in animal herds, but mainly due to the measures (including containment, monitoring, vaccines). The economic costs in Germany from the outbreak starting in 2006 until the re-emergence in 2018 were estimated at approximately 180 million euros (Gethmann et al. 2020). Regardless of the low case numbers, even a single detection is accompanied by significant restrictions, such as the establishment of a **restricted zone** of 150 km, which must be maintained for 2 years. In this zone, all susceptible livestock (including private ones) must be reported and transport of susceptible animals out of the restricted zone is prohibited (Landesuntersuchungsamt Rheinland-Pfalz 2021).

More than 20 years ago, it was suspected that only one **species of biting midges**. *Culicoides imitans*, could transmit bluetongue **virus**, as European outbreaks prior to 1998 only ever occurred in regions where this biting midge species was resident (Wilson and Mellor 2009). The picture changed with the greater spread of **bluetongue** in Europe. For example, in a biting midge monitoring survey of nearly 100 capture sites during the 2006–2007 outbreak in Germany, biting midge species of the *Culicoides obsoletus complex* were captured in ¾ of the cases, but never *Culicoides imitans* (Mehlhorn et al. 2009). Approximately 50 biting midge species are now thought to have transmission potential (Wilson and Mellor 2009). In some European outbreaks, the disease spread first to cattle herds and later to sheep herds, probably due to the greater preference of **biting midges** for cattle (Jacquot et al. 2017). Similar to **Schmallenberg virus**, bluetongue virus replication in biting midges is highly temperature dependent. Below 12 °C, no virus replication occurs; at 15 °C, the **extrinsic incubation period** lasts several weeks, to a few days at 30 °C (Wilson and Mellor 2009). The extrinsic incubation period is the time from when a mosquito ingests a virus until the mosquito can transmit the virus. However, temperature is not considered the most important factor for spread, but rather the density of animal herds and the rapid dispersal of the midges and their population densities (Jacquot et al. 2017).

# Human Arboviruses in Europe: Chikungunya, Dengue and West Nile Viruses

6

## 6.1    Chikungunya Virus in Italy

The **Chikungunya virus** (CHIKV) was first identified in Tanzania in Africa in 1952. It belongs to the alphaviruses in the family *Togaviridae* and is also transmitted as an **arbovirus** by mosquitoes. The **Egyptian tiger mosquito** (*Aedes aegypti*) plays the most important role as a vector, but other species such as the **Asian tiger mosquito** (*Aedes albopictus*) can also transmit the pathogen. Like the Schmallenberg virus, it is an enveloped RNA virus. It not only circulates between mosquitoes and humans, but can also circulate in so-called sylvatic cycles between mosquitoes and animals and then be transmitted to humans. The so-called **Chikungunya fever** lasts one to two weeks and is accompanied by fever, headache, rash and aching limbs. The pain in the limbs can also last for months. In addition, other symptoms such as neurological disorders have come into focus in recent years. The disease is associated with meningitis or Guillian-Barré syndrome in some sufferers (Mehta et al. 2018).

Chikungunya virus is primarily found in Africa, where it causes large epidemics at intervals of 7–20 years (Caglioti et al. 2013). Molecular analysis of the phylogenetic evolution of the virus suggests that it spread from Africa to South and Southeast Asia approximately 100 years ago (Weaver et al. 2018). An **epidemic** that began in Kenya in 2004 and spread to the Indian Ocean islands received particular attention. From here and other regions, many vacationers brought the virus to Europe and the Americas (Caglioti et al. 2013). The first outbreak in Europe occurred in Italy in 2007. In such events, an important question is: was the infectious event introduced only by travellers or triggered by local transmission? In the first case, for example, a traveller brings the disease back home. After this person has recovered, the infection event is over until the next person brings the disease with him. In contrast, **local transmission** is the more dangerous event. Here, local mosquitoes pick up the virus while feeding on the blood of the infected traveler and transmit it to other people. This is called autochthonous transmission. The outbreak in Italy in 2007 was such a case. The first patient, also called an **index patient**, was a man from India who traveled to an Italian village with an infection for a brief visit to his relatives. A short time later, there was an **outbreak** with over 200 proven ill people in two villages, as well as some more distant follow-up cases. The investigation revealed a particularly high density of the **Asian tiger mosquito** in the region, which was responsible for the rapid spread. The outbreak was brought under control by mosquito control (Rezza 2018).

Ten years later, there was another **outbreak** (Marano et al. 2017). In contrast to the 2007 outbreak, the 2017 outbreak initially developed undetected in the summer and only later became conspicuous (Rezza 2018). This resulted in some so-called **clusters**, or local clustering of cases, in different regions (Vairo et al. 2018). In total, about 240 cases were confirmed in the Lazio region. This outbreak was also probably due to a single entry event, as molecular analysis of the virus sequence showed the highest similarity to virus isolates found in India and Pakistan in 2016 (Cella et al. 2018). However, Italy is not the only European country affected so far. With a lower magnitude, some local transmissions of **chikungunya viruses** occurred in France (Grandadam et al. 2011).

These examples illustrate the risk posed by individual travellers who fall ill. However, this also shows that currently the environmental factors must be favourable for the virus, as no local transmission of the virus has been detected in Italy in the intervening years. Suitable **vectors** in high density are needed, as a sick person who is not bitten by a mosquito cannot cause an outbreak. A risk in such cases is the late detection of early virus circulation. In the second case in 2017, viral circulation went undetected for some time before the first clinical cases were detected. In such cases, it is even more difficult to reconstruct the exact course of events. This shows parallels with the early spread of **COVID-19** in northern Italy. **SARS-CoV-2** does not require vectors for transmission, but again spread undetected by direct human-to-human transmission in February 2020. By the time this was recognised as a serious health problem, the level of infestation was so high that shortly afterwards the health system was overloaded.

Many factors play a role in the question of whether larger epidemics are also possible. These include differences in the transmission capacity of different **virus genotypes,** the geographical distribution of mosquitoes, mosquito density, the mobility of infected persons, population density and also temperature. The replication of the virus in the **Asian tiger mosquito** is temperature dependent. In an experimental study, the virus appeared in the mosquitoes' saliva after a few days at 28 °C. At cooler temperatures of 18 °C, it took much longer for the virus to be detectable in saliva (Wimalasiri-Yapa et al. 2019). This is one of the reasons why certain tropical diseases are currently gaining a poor foothold, despite the fact that the pathogens are constantly being introduced into Europe by travellers.

In simulations, the risk for Germany in the near future is relatively low, but the risk for local **chikungunya virus transmission** in France and the Benelux countries in the first half of the twenty-first century is estimated to be higher (Fischer et al. 2013). However, such simulations and risk assessments depend on certain assumptions about factors that can be poorly predicted. For example, it is not yet certain how much the climate change will continue in the coming years and decades and how much the transmitting mosquitoes will spread. In contrast to the simulation, the European disease control agency **ECDC** rates the risk for **chikungunya fever** in Europe higher, due to the high travel activity, the presence of transmissible mosquitoes and the lack of immunity of the population (ECDC 2014). For this reason, **monitoring programmes** should be established or intensified in order to detect the early circulation of the virus in mosquitoes and thus detect local outbreaks in the early phase and prevent epidemics (Vairo et al. 2018).

Efforts to develop **vaccines** have also been underway for some time. One candidate is based on so-called virus-like particles (Chen et al. 2020). These are particles that mimic the surface of the virus in parts without containing the viral genome and thus have no ability to replicate (Vogel 2021). One candidate was safe and immunogenic in phase I/II clinical trials (Chen et al. 2020), and it is hoped that the further hurdles will also be overcome.

## 6.2  Dengue Virus in France

**Dengue fever** is one of the once tropical diseases that is spreading at a worrying rate (Vogel and Schaub 2021). Dengue virus is the most common **arbovirus** in the world and belongs to the flaviviruses. The global spread of the **Asian tiger mosquito** (*Aedes albopictus*) is considered one of the greatest threats regarding the further increase of dengue fever illnesses (Lambrechts et al. 2010). **Dengue virus** is an enveloped virus that contains a single-stranded RNA as the viral genome. There are 4 serotypes, some of which differ in virulence, with DENV-2 as the most dangerous serotype. The first records of epidemics date from Asia, Africa and North America from 1779 to 1780 (Gubler and Clark 1995). Currently, it is estimated that approximately 4 billion people live in risk areas.

From a European perspective, **dengue fever** is currently still a predominantly imported infectious disease, brought to Europe by travellers but with little foothold here. When 130 travellers with fever were tested for dengue fever in Spain, about half were dengue positive. This is not yet worrying, as dengue fever is very common and occurs in many popular holiday destinations. However, dengue **virus** was also found in mosquitoes (**Asian tiger mosquito**) in one patient's home after he returned home. Although no secondary infections resulted from this, it shows that local transmission of dengue fever is possible in principle, provided that transmissible mosquitoes are present (Aranda et al. 2018).

A well-described small outbreak of **dengue fever** with local transmission occurred in France in 2015. In previous years, there were isolated cases of local transmission in France. In 2015, similar to the outbreaks of **chikungunya fever** in Italy, the infections originated from an infected person who had entered the country. On July 4, that person developed a fever. **Secondary infections** transmitted by **Asian tiger mosquitoes** (*A. albopictus*) did not occur until August and extended into September (Succo et al. 2016). This lag phase is not unusual. To the extent that arboviruses are ingested by mosquitoes while sucking blood, they must first develop in the mosquito before the mosquito can transmit the virus via saliva while sucking blood. This **extrinsic incubation period** lasts 1–2 weeks. In the first step, the virus infects the epithelial cells of the mosquito's midgut. In these, the virus multiplies and infects further cells. After 7–10 days, large parts of the midgut are infected and virus is already found in the mosquitoes' salivary glands, where virus replication occurs again (Salazar et al. 2007; Raquin and Lambrechts 2017). The mosquitoes inject the saliva during the blood meal, partly to suppress the coagulation of the ingested blood, and thus transmit the viruses. After the dormant period of about 1 month, another 7 cases then occurred in the region. There was a very intensive response to the outbreak, with surveys of surrounding households, testing, evaluation of information from medical practices in the wider area on fever cases, and intensive control of the mosquito population using insecticides, up to and including removal of possible breeding sites (Succo et al. 2016). As outbreaks also leave a certain number of cases undetected, an **antibody study** was subsequently conducted in the surrounding area, and participants were positive in only a few individual cases. Interestingly, however, approximately 1% of participants tested positive for **West Nile virus**, discussed in the next section (Succo et al. 2018). An important question is the cause of the difference in imported cases for one virus but local transmission by mosquitoes for others. Analysis of data from 2010 to 2018 concluded that the outbreaks were associated with delayed reporting of cases to local health departments, i.e. failure to prevent a possible **chain of infection** through reporting and quarantines (Jourdain et al. 2020).

## 6.3     West Nile Virus in Europe

**West Nile virus** was first isolated in Uganda in 1937 from the blood of a person with fever, so it has been known for a long time. This RNA virus from the flavivirus group circulates between mosquitoes and birds and, with few exceptions, is distributed throughout the world (Kramer et al. 2019; Fig. 6.1). The disease, **West Nile fever,** is a notifiable animal disease. Unlike dengue or chikungunya fever, the main vectors in this case are not *Aedes* mosquitoes, but mosquitoes of the genus *Culex* (Pierson and Diamond 2020). The so-called common house mosquito *Culex pipiens,* which is widespread in Germany and Europe, also belongs to this genus.

Some bird species can become severely ill with West Nile fever, but the infection is inconspicuous in most bird species (Gossner et al. 2017). However, the virus can cause severe disease in horses and humans (Fig. 6.2). In this context, these hosts are so-called **dead-end hosts**, i.e. the chain of infection ends in horses or humans, as infected individuals do not develop sufficient viremia, so that the amount of virus in the blood is not sufficient to be transmitted to mosquitoes again. Humans usually remain asymptomatic (about 80%), but in a few cases (1%) fatal meningitis may develop. People with survived infection often suffer from various long-term sequelae such as headache, weakness or memory loss (Rossi et al. 2010). When horses are infected, approximately 10% of the animals develop neurological symptoms (Gossner et al. 2017).

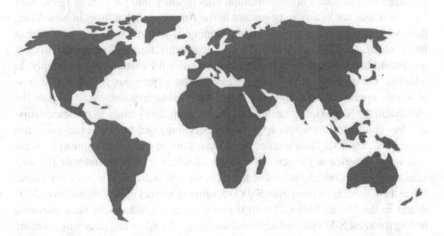

**Fig. 6.1** Distribution area of West Nile virus (red: present; blue: absent). (Image source: Adobe Stock, file no.: 220764030, modified)

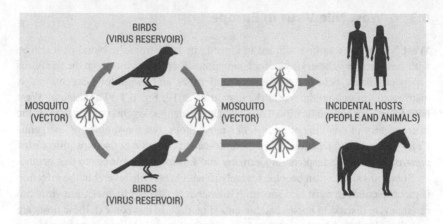

**Fig. 6.2** Transmission cycle of West Nile virus between mosquitoes and birds with occasional infection of humans or horses. (Image source: Adobe Stock, file no.: 220764030, modified)

After the discovery of **West Nile virus**, numerous outbreaks occurred worldwide in the 1950s–1970s, including in Europe (Kramer et al. 2019). After that, things became quiet and the virus was detected only sporadically, including, for example, in smaller outbreaks in Europe involving horses and humans (Napp et al. 2018). The pathogen experienced a brilliant triumph after its entry into the U.S. in 1999. West Nile virus was not known to be present in the Americas at that time. In New York, there were several patients with atypical meningitis within a short period of time in late summer. In addition, dead birds were frequently found. This particularly hot and humid late summer brought ideal opportunities for mosquitoes to multiply. In addition, the manifestation of the virus allowed a previously immigrated new mosquito species, *Culex pipiens*. There are various hypotheses as to how the virus came to New York, from introduction by migratory birds to the importation of virus-containing mosquito eggs in travelers' luggage, but the exact route has never been clarified. The events of New York allowed the virus to spread to North and South America in the years that followed, where it became **endemic** (Kramer et al. 2019). Genetic adaptations of the virus were also detected in the early phase. Since then, there have been over 50,000 confirmed human infections and over 2000 deaths in the United States. Due to the new threat, there has also been increased testing for West Nile virus during blood screening. Based on this screening, as many as 2–four million human infections were estimated in the U.S. by 2010 (Pierson and

Diamond 2020). However, the virus is also dangerous for horses that are exposed to mosquitoes on pastures.

Such events give rise to the legitimate question: can West Nile virus also become endemic in our country? Here in Europe, **West Nile virus** was hardly an issue until the early twenty-first century. An exception was an outbreak in Bulgaria in 1997 with several hundred human infections with neurological symptoms. Beginning in 2004, a new genetic lineage of West Nile virus (lineage 2), first found in Hungary, increasingly spread. From this, **West Nile fever** was found in other European areas in subsequent years (Napp et al. 2018). Data on the occurrence of West Nile virus in different European countries are collected through a reporting system of the European Commission and publicly summarized on a regular basis by the European Centre for Disease Control (**ECDC**) (Bakonyi and Haussig 2020).

In Europe, Greece has been comparatively affected by **West Nile fever** since 2010, with one hundred or over one hundred human infections per year (Rizzoli et al. 2015; Gossner et al. 2017; Bakonyi and Haussig 2020). In 2018, there were over 2000 human infections with **West Nile virus** in the European region and neighbouring countries, with a significant increase compared to the previous year, probably caused by climatic conditions including long periods of higher temperatures (Michel et al. 2019). Fewer cases have been reported in Europe in the two years since. However, in addition to the increasing geographical spread, it is becoming apparent that cases are occurring in successive periods in the same areas, so the virus is probably not repeatedly re-introduced but persists (Bakonyi and Haussig 2020). Interesting insights on this were revealed by monitoring overwintering mosquito species, in particular *Culex pipiens*, in the Czech Republic. While the analysed mosquito populations were negative for WNV in the winter period until 2016, West Nile virus was found in hibernating mosquitoes for the first time in two locations in 2017 (Rudolf et al. 2017).

In Germany, there were no confirmed cases of **West Nile virus infection** in birds before 2018. However, seropositive migratory birds were found in eastern Germany (Michel et al. 2019). The first cases of West Nile virus detection were wild birds, probably due to a single entry event from the Czech Republic (Ziegler et al. 2019). There are several reasons why a cluster of infection occurs in particularly warm summer months. In addition to human behaviors of being more likely to leave windows open during intense heat and spending more time outside until dusk lightly clothed, there are other reasons. Long extended periods of high temperature also affect mosquitoes, which bite more frequently under these conditions. In addition, the virus also develops better in the mosquito. The **extrinsic incubation period** already described changes at higher temperatures. Based on theoretical calculations, it was estimated that West Nile virus would have required a development time of about 3 weeks in many regions of Germany from the 1980s to 2017, but only about 2 weeks in some regions in 2018 (Ziegler et al. 2019). Overall, an increasing spread was found this year, as the virus was sporadically detected in wild birds, horses or humans in different regions such as Berlin, Saxony-Anhalt, Bavaria and Mecklenburg-Western Pomerania (Johnson et al. 2018).

Recently, there were several cases of severe human illness in Leipzig in the late summer of 2020. It is suspected that **West Nile virus** will show up more frequently in this region (Bakonyi and Haussig 2020). Overall, however, we must also be prepared for West Nile virus to also appear more widely in Germany if environmental conditions are favourable with extended periods of heat. Due to the strong spread of the virus and the increasing risk of becoming **endemic** also in new areas and then causing a high number of infections in humans, **vaccines** are a good alternative to be prepared for such a scenario. Vaccine development against **West Nile fever** now has a 20-year history. While some vaccines are available for horses, none are yet available for humans (Kaiser and Barrett 2019).

# Summary

Germany and other European countries had until recently and predominantly still have the charm that in this temperate zone we do not have to worry much about exotic diseases in everyday life. Even insects as carriers need not yet be feared. Relatively common are only in some risk areas certain **infectious diseases** transmitted by ticks. But not only diseases transmitted by insects are rare in Germany. We only know infectious diseases such as SARS or Ebola fever from television. There are many factors that influence the spread of existing infectious diseases and the introduction of new ones. **Globalization** contributes to the large-scale transport of commercial goods, food and animals, and the fact that millions of people travel to different countries several times a year, bringing unwanted "passengers" with them, in the form of mosquitoes or their eggs, contaminated food or viruses in the case of fresh infections. In addition, the **growing population** of the **earth** has led to more and larger urban areas, a sharp increase and enlargement of livestock populations and the progressive invasion of wildlife habitats. Furthermore, steadily advancing **global warming** is promoting the invasion of previously temperate zones by new pathogens and insects. The precursors are already being noticed as new mosquito species are settling themselves that can be efficient vectors. The risk of regional transmission of tropical arboviruses will still be low in the next few years, but it will also depend on whether we again experience extreme warm periods in summer in the next few years, as we have in some cases in recent years. The example of the **West Nile virus** in the USA also proves that sometimes temporary extreme conditions or coincidences can lead to new viruses persisting and experiencing a triumphal procession in new areas.

After the victories against opponents such as **SARS** or **Ebola**, it seemed safe to bring new dangers under control with the classical means of disease control. COVID-19, with its ambivalent mixture of high lethality and a large proportion of

© The Author(s), under exclusive license to Springer Fachmedien Wiesbaden
GmbH, part of Springer Nature 2023
P. U. B. Vogel, G. A. Schaub, *New Infectious Diseases in Germany
and Europe*, essentials, https://doi.org/10.1007/978-3-658-41826-7_7

asymptomatic cases, has shown how difficult it is to fight invisible opponents (=asymptomatic carriers) that undermined early containment measures. In addition, the pathogen exhibits high infectivity. These factors have made COVID-19 one of the worst **pandemics** in human history from a health, social and economic perspective. However, we should not be lulled into the belief that we will be safe from coronaviruses for years to come. The processes that lead to the accumulation of **spillover events** do not remain constant, but are increasing.

As different as the possible **infectious diseases** are, as different are measures for early detection or prevention. While for the detection of certain **arboviruses** we need further intensification of existing monitoring of mosquito populations and wildlife, for avian influenza sentinel animals are a good way to detect the circulation of new influenza A strains. In contrast, to prevent infectious diseases such as **COVID-19,** one must leave one's own soil and focus on the source. Regular and widespread serum monitoring of hotspot workers (e.g., animal markets) for new coronavirus variants could help detect early circulation of dangerous new variants. Even in the case of **SARS,** there does not seem to have been "the" first **index patient** that started the pandemic, as there were workers from animal markets a year earlier who had antibodies to SARS or SARS-like viruses in their blood. Especially for zoonotic diseases that go through a months- or years-long trial phase of **spillover**, proactive monitoring could help nip new zoonotic diseases "in the bud." But that would still be mitigation, allowing it to happen first, then stopping it early. True prevention would involve, for example, creating alternatives for the ongoing expansion of livestock farming into wildlife habitats, which is anything but easy.

# What Readers Can Take Away from This *Essential*

- Changing conditions favour the spread of subtropical and tropical infectious diseases also in temperate zones
- New mosquito species with the ability to transmit certain pathogens are increasingly settling in our areas
- The risk in Germany for many genuine tropical diseases is still low, but will continue to increase in the coming years and decades
- COVID-19 has shown that classical disease control has its limits
- Active and passive monitoring programmes must be intensified in order to be able to react to outbreaks at an early stage

© The Author(s), under exclusive license to Springer Fachmedien Wiesbaden
GmbH, part of Springer Nature 2023
P. U. B. Vogel, G. A. Schaub, *New Infectious Diseases in Germany
and Europe*, essentials, https://doi.org/10.1007/978-3-658-41826-7

# References

Adlhoch C, Gossner C, Koch G et al (2014) Comparing introduction to Europe of highly pathogenic avian influenza viruses A(H5N8) in 2014 and A(H5N1) in 2005. Euro Surveill 19:20996. https://doi.org/10.2807/1560-7917.es2014.19.50.20996

Akiner MM, Öztürk M, Baser AB (2019) Arboviral screening of invasive *Aedes* species in northeastern Turkey: West Nile virus circulation and detection of insect-only viruses. PloS Negl Trop Dis 13:e0007334. https://doi.org/10.1371/journal.pntd.0007334

Aranda C, Martínez MJ, Montalvo T et al (2018) Arbovirus surveillance: first dengue virus detection in local *Aedes albopictus* mosquitoes in Europe, Catalonia, Spain, 2015. Euro Surveill 23(47):1700837. https://doi.org/10.2807/1560-7917

Ayllón T, Nijhof AM, Weiher W et al (2014) Feeding behaviour of *Culicoides* spp. (Diptera: Ceratopogonidae) on cattle and sheep in northeast Germany. Parasit Vectors 7:34. https://doi.org/10.1186/1756-3305-7-34

Bakonyi T, Haussig JM (2020) West Nile virus keeps on moving up in Europe. Euro Surveill 25:2001938. https://doi.org/10.2807/1560-7917.ES.2020.25.46.2001938

Becker N, Jöst A, Weitzel T (2012) The *Culex pipiens* complex in Europe. J Am Mosq Control Assoc 28:53–67. https://doi.org/10.2987/8756-971X-28.4s.53

Biront P, Castryck F, Leunen J (1987) An epizootic of African swine fever in Belgium and its eradication. Vet Rec 120:432–434. https://doi.org/10.1136/vr.120.18.432

Bode L, Xie P, Dietrich DE et al (2020) Are human Borna disease virus 1 infections zoonotic and fatal? Lancet Infect Dis 20:650–651. https://doi.org/10.1016/S1473-3099(20)30380-7

Bonnet SI, Bouhsira E, De Regge N et al (2020) Putative role of arthropod vectors in African swine fever virus transmission in relation to their bio-ecological properties. Viruses 12:778. https://doi.org/10.3390/v12070778

Bouvier NM, Palese P (2008) The biology of influenza viruses. Vaccine 4:D49–D53. https://doi.org/10.1016/j.vaccine.2008.07.039

Bundesumweltamt (2019) Asiatische Tigermücke. https://www.umweltbundesamt.de/asiatische-tigermuecke#alternative-bekampfungsmassnahmen. Accessed on: 12. Febr. 2021

Caglioti C, Lalle E, Castilletti C et al (2013) Chikungunya virus infection: an overview. New Microbiol 36:211–227

Cella E, Riva E, Salemi M et al (2018) The new chikungunya virus outbreak in Italy possibly originated from a single introduction from Asia. Pathog Glob Health 112:93–95. https://doi.org/10.1080/20477724.2017.1406565

Chen GL, Coates EE, Plummer SH et al (2020) Effect of a chikungunya virus-like particle vaccine on safety and tolerability outcomes: a randomized clinical trial. J Am Med Ass 323:1369–1377. https://doi.org/10.1001/jama.2020.2477

Chenais E, Depner K, Guberti V et al (2019) Epidemiological considerations on African swine fever in Europe 2014–2018. Porcine Health Manag 5:6. https://doi.org/10.1186/s40813-018-0109-2

Chitimia-Dobler L, Schaper S, Rieß R et al (2019) Imported *Hyalomma* ticks in Germany in 2018. Parasit Vectors 12:134. https://doi.org/10.1186/s13071-019-3380-4

Claine F, Coupeau D, Wiggers L et al (2015) Schmallenberg virus infection of ruminants: challenges and opportunities for veterinarians. Vet Med (auckl) 6:261–272. https://doi.org/10.2147/VMRR.S83594

Collins ÁB, Doherty ML, Barrett DJ et al (2019) Schmallenberg virus: a systematic international literature review (2011–2019) from an Irish perspective. Ir Vet J 72:9. https://doi.org/10.1186/s13620-019-0147-3

Conraths FJ, Gethmann JM, Staubach C et al (2009) Epidemiology of bluetongue virus serotype 8, Germany. Emerg Infect Dis 15:433–435. https://doi.org/10.3201/eid1503.081210

Corman VM, Muth D, Niemeyer D et al (2018) Hosts and sources of endemic human coronaviruses. Adv Virus Res 100:163–188. https://doi.org/10.1016/bs.aivir.2018.01.001

CSSE (2021) Coronavirus 2019-nCoV global cases by Johns Hopkins CSSE. https://coronavirus.jhu.edu/map.html. Accessed on: 18. Mai 2021

Cwynar P, Stojkov J, Wlazlak K (2019) African swine fever status in Europe. Viruses 11:310. https://doi.org/10.3390/v11040310

Drosten C, Günther S, Preiser W et al (2003) Identification of a novel coronavirus in patients with severe acute respiratory syndrome. N Engl J Med 348:1967–1976. https://doi.org/10.1056/NEJMoa030747

Dunphy BM, Kovach KB, Gehrke EJ et al (2019) Long-term surveillance defines spatial and temporal patterns implicating *Culex tarsalis* as the primary vector of West Nile virus. Sci Rep Apr 9:6637. https://doi.org/10.1038/s41598-019-43246-y

ECDC (2014) Risk assessment for chikungunya in the EU continental and overseas countries, territories and departments. https://www.ecdc.europa.eu/en/chikungunya/threats-and-outbreaks/risk-assessment-chikungunya-eu#:~:text=The%20risk%20of%20chikungunya%20fever,Mediterranean%20coast)%20and%20population%20susceptibility. Accessed on: 5. März 2021

Endalew AD, Faburay B, Wilson WC et al (2019) Schmallenberg disease – a newly emerged *Culicoides*-borne viral disease of ruminants. Viruses 11:1065. https://doi.org/10.3390/v11111065

European Food Safety Authority, European Centre for Disease Prevention and Control and European Union Reference Laboratory for Avian Influenza, , Adlhoch C et al (2020) Avian influenza overview February – May 2020. EFSA J 18:e06194. https://doi.org/10.2903/j.efsa.2020.6194

Fischer D, Thomas SM, Suk JE et al (2013) Climate change effects on Chikungunya transmission in Europe: geospatial analysis of vector's climatic suitability and virus' temperature requirements. Int J Health Geogr 12:51. https://doi.org/10.1186/1476-072X-12-51

FLI (2021a) Blauzungenkrankheit (BT). https://www.fli.de/de/aktuelles/tierseuchengeschehen/blauzungenkrankheit/. Accessed on: 9. März 2021a

FLI (2021b) Afrikanische Schweinepest. https://www.fli.de/de/aktuelles/tierseuchengeschehen/afrikanische-schweinepest/. Accessed on: 10. März 2021b

Früh L, Kampen H, Koban MB et al (2020) Oviposition of Aedes japonicus japonicus (Diptera: Culicidae) and associated native species in relation to season, temperature and land use in western Germany. Parasit Vectors 13:623. https://doi.org/10.1186/s13071-020-04461-z

Gethmann J, Probst C, Conraths FJ (2020) Economic impact of a bluetongue serotype 8 epidemic in Germany. Front Vet Sci 7:65. https://doi.org/10.3389/fvets.2020.00065

Gillim-Ross L, Subbarao K (2006) Emerging respiratory viruses: challenges and vaccine strategies. Clin Microbiol Rev 19:614–636. https://doi.org/10.1128/CMR.00005-06

Globig A, Staubach C, Sauter-Louis C et al (2018) Highly pathogenic avian influenza H5N8 clade 2.3.4.4b in Germany in 2016/2017. Front Vet Sci 4:240. https://doi.org/10.3389/fvets.2017.00240

Gossner CM, Marrama L, Carson M et al (2017) West Nile virus surveillance in Europe: moving towards an integrated animal-human-vector approach. Euro Surveill 22:30526. https://doi.org/10.2807/1560-7917.ES.2017.22.18.30526

Graham RL, Donaldson EF, Baric RS (2013) A decade after SARS: strategies for controlling emerging coronaviruses. Nat Rev Microbiol 11:836–848. https://doi.org/10.1038/nrmicro3143

Grandadam M, Caro V, Plumet S et al (2011) Chikungunya virus, southeastern France. Emerg Infect Dis 17:910–913. https://doi.org/10.3201/eid1705.101873

Greenberg SB (2016) Update on human rhinovirus and coronavirus infections. Semin Respir Crit Care Med 37:555–571. https://doi.org/10.1055/s-0036-1584797

Gubler DJ, Clark GG (1995) Dengue/dengue hemorrhagic fever: the emergence of a global health problem. Emerg Infect Dis 1:55–57

Guinat C, Gogin A, Blome S et al (2016) Transmission routes of African swine fever virus to domestic pigs: current knowledge and future research directions. Vet Rec 178:262–267. https://doi.org/10.1136/vr.103593

Guo YR, Cao QD, Hong ZS et al (2020) The origin, transmission and clinical therapies on coronavirus disease 2019 (COVID-19) outbreak – an update on the status. Mil Med Res 7: 11. https://doi.org/10.1186/s40779-020-00240-0

Hagenaars TJ, Backx A, van Rooij EMA et al (2021) Within-farm transmission characteristics of bluetongue virus serotype 8 in cattle and sheep in the Netherlands, 2007–2008. PloS One 16:e0246565. https://doi.org/10.1371/journal.pone.0246565

Harapan H, Itoh N, Yufika A et al (2020) Coronavirus disease 2019 (COVID-19): a literature review. J Infect Public Health 13(5):667–673. https://doi.org/10.1016/j.jiph.2020.03.019

Hemmer CJ, Emmerich P, Loebermann M et al (2018) Mücken und Zecken als Krankheitsvektoren: der Einfluss der Klimaerwärmung. Dtsch Med Wochenschr 143: 1714–1722. https://doi.org/10.1055/a-0653-6333

Hilbe M, Herrsche R, Kolodziejek J et al (2006) Shrews as reservoir hosts of Borna disease virus. Emerg Infect Dis 12:675–677. https://doi.org/10.3201/eid1204.051418

Holicki CM, Scheuch DE, Ziegler U et al (2020) German *Culex pipiens* biotype *molestus* and *Culex torrentium* are vector-competent for Usutu virus. Parasit Vectors 13:625. https://doi.org/10.1186/s13071-020-04532-1

Jacquot M, Nomikou K, Palmarini M et al (2017) Bluetongue virus spread in Europe is a consequence of climatic, landscape and vertebrate host factors as revealed by phylogeographic inference. Proc Biol Sci 284:20170919. https://doi.org/10.1098/rspb.2017.0919

Jansen S, Heitmann A, Lühken R et al (2019) *Culex torrentium*: a potent vector for the transmission of West Nile Virus in central Europe. Viruses 11(6):E492. https://doi.org/10.3390/v11060492

Johnson N, Fernández de Marco M, Giovannini A et al (2018) Emerging mosquito-borne threats and the response from European and Eastern Mediterranean countries. Int J Environ Res Public Health 15:2775. https://doi.org/10.3390/ijerph15122775

Jourdain F, Roiz D, de Valk H et al (2020) From importation to autochthonous transmission: drivers of chikungunya and dengue emergence in a temperate area. PloS Negl Trop Dis 14:e0008320. https://doi.org/10.1371/journal.pntd.0008320

Kahn JS, McIntosh K (2005) History and recent advances in coronavirus discovery. Pediatr Infect Dis J 24:223–227. https://doi.org/10.1097/01.inf.0000188166.17324.60

Kaiser JA, Barrett ADT (2019) Twenty years of progress toward West Nil virus vaccine development. Viruses 11:823. https://doi.org/10.3390/v11090823

Kameke D, Kampen H, Walther D (2017) Acitivity of *Culicoides* spp. (Diptera: Ceratopogonidae) inside and outside of livestock stables in late winter and spring. Parasitol Res 116:881–889. https://doi.org/10.1007/s00436-016-5361-2

Kampen H, Kuhlisch C, Fröhlich A et al (2016) Occurence and spread of the invasive asian bush mosquito *Aedes japonicus japonicus* (Diptera: Culicidae) in West and North Germany since detection in 2012 and 2013, respectively. PloS One 11:e0167948. https://doi.org/10.1371/journal.pone.0167948

Karch H, Denamur E, Dobrindt U et al (2012) The enemy within us: lessons from the 2011 European *Escherichia coli* O104:H4 outbreak. EMBO Mol Med 4:841–848. https://doi.org/10.1002/emmm.201201662

Karger A, Pérez-Núñez D, Urquiza J et al (2019) An update on African swine fever virology. Viruses 11:864. https://doi.org/10.3390/v11090864

Kaufmann C, Schaffner F, Mathis A (2009) Monitoring of biting midges (*Culicoides* spp.), the potential vectors of the bluetongue virus, in the 12 climatic regions of Switzerland. Schweiz Arch Tierheilkd 151:205–213. https://doi.org/10.1024/0036-7281.151.5.205

Klein H, Asseo K, Karni N et al (2021) Onset, duration and unresolved symptoms, including smell and taste changes, in mild COVID-19 infection: a cohort study in Israeli patients. Clin Microbiol Infect 16:S1198-743X(21)00083-5. https://doi.org/10.1016/j.cmi.2021.02.008

Koban MB, Kampen H, Scheuch DE et al (2019) The Asian bush mosquito *Aedes japonicus japonicus* (Diptera: Culicidae) in Europe, 17 years after its first detection, with a focus on monitoring methods. Parasit Vectors 12:109. https://doi.org/10.1186/s13071-019-3349-3

Koenraadt CJM, Möhlmann TWR, Verhulst NO et al (2019) Effect of overwintering on survival and vector competence of the West Nile virus vector *Culex pipiens*. Parasit Vectors 12:147. https://doi.org/10.1186/s13071-019-3400-4

Kramer LD, Ciota AT, Kilpatrick AM (2019) Introduction, spread, and establishment of West Nile virus in the Americas. J Med Entomol 56:1448–1455. https://doi.org/10.1093/jme/tjz151

Kuhn JH, Bavari S (2017) Asymptomatic Ebola virus infections – myth or reality? Lancet Infect Dis 17:570–571. https://doi.org/10.1016/S1473-3099(17)30110-X

Kupferschmidt K (2012) Neue Seuche im Stall, alte Seuchen, neues Virus, ständige Gefahr. Potsdamer Neueste Nachrichten. https://www.pnn.de/wissenschaft/ueberregional/neue-seuche-im-stall-alte-seuchen-neues-virus-staendige-gefahr/21874470.html. Accessed on: 3. März 2021

Kupke A, Becker S, Wewetzer K et als (2019) Intranasal Borna disease virus (BoDV-1) infection: insights into initial steps and potential contagiosity. Int J Mol Sci 20:1318. https://doi.org/10.3390/ijms20061318

Lambrechts L, Scott TW, Gubler DJ (2010) Consequences of the expanding global distribution of *Aedes albopictus* for dengue virus tranmission. PLoS Negl Trop Dis 4:e646. https://doi.org/10.1371/journal.pntd.0000646

Landesuntersuchungsamt Rheinland-Pfalz (2021) Blauzungenkrankheit. https://lua.rlp.de/de/unsere-themen/lexikon/lexikon-b/blauzungenkrankheit/. Accessed on: 14. März 2021

Lecollinet S, Pronost S, Coulpier M et al (2019) Viral equine encephalitis, a growing threat to the horse population in Europe? Viruses 12:23. https://doi.org/10.3390/v12010023

Leggewie M, Badusche M, Rudolf M et al (2016) *Culex pipiens* and *Culex torrentium* populations from Central Europe are susceptible to West Nile virus infection. One Health 2:88–94. https://doi.org/10.1016/j.onehlt.2016.04.001

LGL (2021) Schmallenberg-Virus (Europäisches Shamonda-like Orthobunyavirus). https://www.lgl.bayern.de/tiergesundheit/tierkrankheiten/virusinfektionen/schmallenberg_virus/. Accessed on: 3. März 2021

Ludwig H, Bode L (2000) Borna disease virus: new aspects on infection, disease, diagnosis and epidemiology. Rev Sci Tech 19:259–288. https://doi.org/10.20506/rst.19.1.1217

Marano G, Pupella S, Pati I et al (2017) Ten years since the last Chikungunya virus outbreak in Italy: history repeats itself. Blood Transfus 15:489–490. https://doi.org/10.2450/2017.0215-17

Martinet JP, Ferté H, Failloux AB et al (2019) Mosquitoes of North-Western Europe as potential vectors of arboviruses: a review. Viruses 11:1059. https://doi.org/10.3390/v11111059

Mazur-Panasiuk N, Żmudzki J, Woźniakowski G (2019) African swine fever virus – persistence in different environmental conditions and the possibility of its indirect transmission. J Vet Res 63:303–310. https://doi.org/10.2478/jvetres-2019-0058

Maxmen A (2021) WHO report into COVID pandemic origins zeroes in on animal markets, not labs. Nature 592:173–174

McMillan JR, Marcet PL, Hoover CM et al (2019) Feeding success and host selection by *Culex quinquefasciatus* Say mosquitoes in experimental trials. Vector Borne Zoonotic Dis 19:540–548. https://doi.org/10.1089/vbz.2018.2381

Mehlhorn H, Walldorf V, Klimpel S et al (2009) Bluetongue disease in Germany (2007-2008): monitoring of entomological aspects. Parasitol Res 105:313–319. https://doi.org/10.1007/s00436-009-1416-y

Mehta R, Gerardin P, de Brito CAA et al (2018) The neurological complications of chikungunya virus: a systematic review. Rev Med Virol 28:e1978. https://doi.org/10.1002/rmv.1978

Michel F, Sieg M, Fischer D et al (2019) Evidence for West Nile virus and Usutu virus infections in wild and resident birds in Germany, 2017 and 2018. Viruses 11:674. https://doi.org/10.3390/v11070674

Morens DM, Daszak P, Taubenberger JK (2020) Escaping Pandora's box – another novel coronavirus. N Engl J Med 382(14):1293–1295. https://doi.org/10.1056/NEJMp2002106

Napp S, Petrić D, Busquets N (2018) West Nile virus and other mosquito-borne viruses present in Eastern Europe. Pathog Glob Health 112:233–248. https://doi.org/10.1080/20477724.2018.1483567

NDR (2021) Geflügelpest im Kreis Plön: 76.000 Hühner werden getötet. https://www.ndr.de/nachrichten/schleswig-holstein/Gefluegelpest-im-Kreis-Ploen-76000-Huehner-werden-getoetet,vogelgrippe584.html. Accessed on: 8. März 2021

Peeri NC, Shrestha N, Rahman MS et al (2020) The SARS, MERS and novel coronavirus (COVID-19) epidemics, the newest and biggest global health threats: what lessons have we learned? Int J Epidemiol 49(3):717–726. https://doi.org/10.1093/ije/dyaa033

Pierson TC, Diamond MS (2020) The continued threat of emerging flaviviruses. Nat Microbiol 5:796–812. https://doi.org/10.1038/s41564-020-0714-0

Pohlmann A, Starick E, Grund C (2018) Swarm incursions of reassortants of highly pathogenic avian influenza virus strains H5N8 and H5N5, clade 2.3.4.4b, Germany, winter 2016/17. Sci Rep 8:15. https://doi.org/10.1038/s41598-017-16936-8

Raquin V, Lambrechts L (2017) Dengue virus replicates and accumulates in *Aedes aegypti* salivary glands. Virology 507:75–81. https://doi.org/10.1016/j.virol.2017.04.009

Reuss F, Wieser A, Niamir A et al (2018) Thermal experiments with the Asian bush mosquito (*Aedes japonicus japonicus*) (Diptera: Culicidae) and implications for its distribution in Germany. Parasit Vectors 11:81. https://doi.org/10.1186/s13071-018-2659-1

Rezza G (2018) Chikungunya is back in Italy: 2007–2017. J Travel Med 25. https://doi.org/10.1093/jtm/tay004

Richt JA, Pfeuffer I, Christ M et al (1997) Borna disease virus infection in animals and humans. Emerg Infect Dis 3:343–352. https://doi.org/10.3201/eid0303.970311

Rizzoli A, Jimenez-Clavero MA, Barzon L et al (2015) The challenge of West Nile virus in Europe: knowledge gaps and research priorities. Euro Surveill 20:21135. https://doi.org/10.2807/1560-7917.es2015.20.20.21135

RKI (2021) RKI zu humanen Erkrankungen mit aviärer Influenza (Vogelgrippe). https://www.rki.de/DE/Content/InfAZ/Z/ZoonotischeInfluenza/Vogelgrippe.html. Accessed on: 8. März 2021

Rossi SL, Ross TM, Evans JD (2010) West Nile virus. Clin Lab Med 30:47–65. https://doi.org/10.1016/j.cll.2009.10.006

Rubbenstroth D, Niller HH, Angstwurm K et al (2020) Are human Borna disease virus 1 infections zoonotic and fatal? – Authors' reply. Lancet Infect Dis 20:651. https://doi.org/10.1016/S1473-3099(20)30379-0

Rudolf I, Betášová L, Blažejová H et al (2017) West Nile virus in overwintering mosquitoes, central Europe. Parasit Vectors 10:452. https://doi.org/10.1186/s13071-017-2399-7

Salazar MI, Richardson JH, Sánchez-Vargas I (2007) Dengue virus type 2: replication and tropisms in orally infected *Aedes aegypti* mosquitoes. BMC Microbiol 30(7):9

Sauter-Louis C, Forth JH, Probst C et al (2020) Joining the club: First detection of African swine fever in wild boar in Germany. Transbound Emerg Dis 68:1744–1752. https://doi.org/10.1111/tbed.13890

Schulz K, Conraths FJ, Blome S et al (2019) African swine fever: fast and furious or slow and steady? Viruses 11:866. https://doi.org/10.3390/v11090866

Sellwood C, Asgari-Jirhandeh N, Salimee S (2007) Bird flu: if or when? Planning for the next pandemic. Postgrad Med J 83:445–450. https://doi.org/10.1136/pgmj.2007.059253

Sick F, Beer M, Kampen H et al (2019) *Culicoides* biting midges – underestimated vectors for arboviruses of public health and veterinary importance. Viruses 11:376. https://doi.org/10.3390/v11040376

Stefano GB (2021) Historical insight into infections and disorders associated with neurological and psychiatric sequelae similar to long COVID. Med Sci Monit 27:e931447. https://doi.org/10.12659/MSM.931447

Succo T, Leparc-Goffart I, Ferré JB et al (2016) Autochthonous dengue outbreak in Nîmes, South of France, July to September 2015. Euro Surveill 26:21. https://doi.org/10.2807/1560-7917.ES.2016.21.21.30240

Succo T, Noël H, Nikolay B et al (2018) Dengue serosurvey after a 2-month long outbreak in Nîmes, France, 2015: was there more than met the eye? Euro Surveill 23:1700482. https://doi.org/10.2807/1560-7917.ES.2018.23.23.1700482

Sun Z, Thilakavathy K, Kumar SS (2020) Potential factors influencing repeated SARS outbreaks in China. Int J Environ Res Public Health 17:1633. https://doi.org/10.3390/ijerph17051633

tagesschau.de (2021) China verhängt Importverbot. https://www.tagesschau.de/wirtschaft/schweinepest-brandenburg-import-china-101.html. Accessed on: 10. März 2021

Tappe D, Frank C, Offergeld R (2019) Low prevalence of Borna disease virus 1 (BoDV-1) IgG antibodies in humans from areas endemic for animal Borna disease of Southern Germany. Sci Rep 9:20154. https://doi.org/10.1038/s41598-019-56839-4

Tarlinton R, Daly J, Dunham S et al (2012) The challenge of Schmallenberg virus emergence in Europe. Vet J 194:10–18. https://doi.org/10.1016/j.tvjl.2012.08.017

The Lancet (2020a) Facing up to long COVID. Lancet 396:1861. https://doi.org/10.1016/S0140-6736(20)32662-3

The Lancet (2020b) COVID-19: fighting panic with information. Lancet 395:537. https://doi.org/10.1016/S0140-6736(20)30379-2

Uyeki TM, Peiris M (2019) Novel avian influenza A virus infections of humans. Infect Dis Clin North Am 33:907–932. https://doi.org/10.1016/j.idc.2019.07.003

Vairo F, Pietrantonj CD, Pasqualini C et al (2018) The surveillance of chikungunya virus in a temperate climate: challenges and possible solutions from the experience of Lazio region. Italy. Viruses 10:501. https://doi.org/10.3390/v10090501

Verhagen JH, Fouchier RAM, Lewis N (2021) Highly pathogenic avian influenza viruses at the wild-domestic bird interface in Europe: future directions for research and surveillance. Viruses 13:212. https://doi.org/10.3390/v13020212

Vogel PUB (2021) COVID-19: Suche nach einem Impfstoff, 2. Springer Spektrum, Wiesbaden, Aufl. https://doi.org/10.1007/978-3-658-31340-1

Vogel PUB, Schaub GA (2021) Seuchen, alte und neue Gefahren – Von der Pest bis COVID-19. Springer Spektrum, Wiesbaden. https://doi.org/10.1007/978-3-658-32953-2

Vogels CBF, Fros JJ, Göertz GP et al (2016) Vector competence of northern European *Culex pipiens* biotypes and hybrids for West Nile virus is differentially affected by temperature. Parasit Vectors 9:393. https://doi.org/10.1186/s13071-016-1677-0

Walther D, Kampen H (2017) The citizen science project 'Mueckenatlas' helps monitor the distribution and spread of invasive mosquito species in Germany. J Med Entomol 54: 1790–1794. https://doi.org/10.1093/jme/tjx166

Wang F, Kream RM, Stefano GB (2020) Long-term respiratory and neurological sequelae of COVID-19. Med Sci Monit 26:e928996. https://doi.org/10.12659/MSM.928996

Weaver SC, Charlier C, Vasilakis N et al (2018) Zika, chikungunya, and other emerging vector-borne viral diseases. Annu Rev Med 69:395–408. https://doi.org/10.1146/annurev-med-050715-105122

Webster RG, Govorkova EA (2014) Continuing challenges in influenza. Ann N Y Acad Sci 1323:115–139. https://doi.org/10.1111/nyas.12462

Weissenböck H, Bagó Z, Kolodziejek J et al (2017) Infections of horses and shrews with Bornaviruses in Upper Austria: a novel endemic area of Borna disease. Emerg Microbes Infect 6:e52. https://doi.org/10.1038/emi.2017.36

Werner D, Groschupp S, Bauer C et al (2020) Breeding habitat preferences of major *Culicoides* species (Diptera: Ceratopogonidae) in Germany. Int J Environ Res Public Health 17:5000. https://doi.org/10.3390/ijerph17145000

Wernike K, Beer M (2020) Schmallenberg virus: to vaccinated, or not to vaccinate? Vaccines (Basel) 8:287. https://doi.org/10.3390/vaccines8020287

WHO (2004) Summary of probable SARS cases with onset of illness from 1 November 2002 to 31 July 2003. https://www.who.int/csr/sars/country/table2004_04_21/en/. Accessed on: 30. März 2021

WHO (2020) MERS situation update. https://www.emro.who.int/health-topics/mers-cov/mers-outbreaks.html. Accessed on: 20. Febr. 2021

Wilder-Smith A, Teleman MD, Heng BH et al (2005) Asymptomatic SARS coronavirus infection among healthcare workers, Singapore. Emerg Infect Dis 11:1142–1145. https://doi.org/10.3201/eid1107.041165

Wilson AJ, Mellor PS (2009) Bluetongue in Europe: past, present and future. Philos Trans R Soc Lond B Biol Sci 364:2669–2681. https://doi.org/10.1098/rstb.2009.0091

Wimalasiri-Yapa BMCR, Stassen L, Hu W et al (2019) Chikungunya virus transmission at low temperature by *Aedes albopictus* mosquitoes. Pathogens 8:149. https://doi.org/10.3390/pathogens8030149

Xu J, Zhao S, Teng T et al (2020) Systematic comparison of two animal-to-human transmitted human coronaviruses: SARS-CoV-2 and SARS-CoV. Viruses 12:E244. https://doi.org/10.3390/v12020244

Ye ZW, Yuan S, Yuen KS et al (2020) Zoonotic origins of human coronaviruses. Int J Biol Sci 16:1686–1697. https://doi.org/10.7150/ijbs.45472

Yuen KS, Ye ZW, Fung SY et al (2020) SARS-CoV-2 and COVID-19: the most important research questions. Cell Biosci 10:40. https://doi.org/10.1186/s13578-020-00404-4

Zaki AM, van Boheemen S, Bestebroer TM et al (2012) Isolation of a novel coronavirus from a man with pneumonia in Saudi Arabia. N Engl J Med 367:1814–1820. https://doi.org/10.1056/NEJMoa1211721

Ziegler U, Lühken R, Keller M et al (2019) West Nile virus epizootic in Germany, 2018. Antiviral Res 162:39–43. https://doi.org/10.1016/j.antiviral.2018.12.005